Test Item File
David R. Curott

COLLEGE
PHYSICS
THIRD EDITION

JERRY D. WILSON / ANTHONY J. BUFFA

PRENTICE HALL, UPPER SADDLE RIVER, NJ 07458

Acquisition Editor: **Wendy Rivers**
Production Editor: ***Ann Marie Longobardo***
Special Projects Manager: ***Barbara A. Murray***
Supplement Cover Manager: ***Paul Gourhan***
Manufacturing Buyer: ***Ben D. Smith***

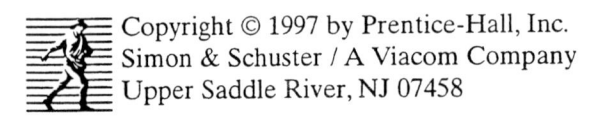

Copyright © 1997 by Prentice-Hall, Inc.
Simon & Schuster / A Viacom Company
Upper Saddle River, NJ 07458

All rights reserved. No part of this book may be
reproduced in any form or by any means,
without permission in writing from the publisher.

Printed in the United States of America

10 9 8 7 6 5 4 3

ISBN 0-13-505124-X

Prentice-Hall International (UK) Limited, *London*
Prentice-Hall of Australia Pty. Limited, *Sydney*
Prentice-Hall Canada Inc., *Toronto*
Prentice Hall Hispanoamericana, S.A., *Mexico*
Prentice-Hall of India Private Limited, *New Delhi*
Prentice-Hall of Japan, Inc., *Tokyo*
Simon & Schuster Asia Pte. Ltd., *Singapore*
Editoria Prentice-Hall do Brasil, *Ltda., Rio de Janero*

Contents

Chapter 1	Units and Problem Solving	1
Chapter 2	Kinematics: Description of Motion	17
Chapter 3	Motion in Two Dimensions	33
Chapter 4	Force and Motion	51
Chapter 5	Work and Energy	69
Chapter 6	Momentum and Collisions	85
Chapter 7	Circular Motion and Gravitation	102
Chapter 8	Rotational Motion and Equilibrium	123
Chapter 9	Solids and Fluids	139
Chapter 10	Temperature	153
Chapter 11	Heat	167
Chapter 12	Thermodynamics	183
Chapter 13	Vibrations and Waves	199
Chapter 14	Sound	216
Chapter 15	Electric Charge, Forces, and Fields	229
Chapter 16	Electric Potential, Energy, and Capacitance	247
Chapter 17	Electric Current and Resistance	260
Chapter 18	Basic Electric Circuits	277
Chapter 19	Magnetism	291
Chapter 20	Electromagnetic Induction	305
Chapter 21	AC Circuits	319
Chapter 22	Geometrical Optics: Reflection and Refraction of Light	333
Chapter 23	Mirrors and Lenses	351
Chapter 24	Physical Optics: The Wave Nature of Light	365
Chapter 25	Optical Instruments	379
Chapter 26	Relativity	392
Chapter 27	Quantum Physics	407
Chapter 28	Quantum Mechanics	418
Chapter 29	The Nucleus	435
Chapter 30	Nuclear Reactions and Elementary Particles	450

Chapter 1

MATCHING

Match the physical unit to the the physical quantity.

a) 1000 g
b) 10^{-9} s
c) 1 in. = 2.54 cm
d) 0.062 yd = 6.2 X 10^{-2} yd
e) 10^6 W

1. kilogram

   Answer: a) 1000 g Difficulty: 2

2. nano-second

   Answer: b) 10^{-9} s Difficulty: 2

3. conversion

   Answer: c) 1 in. = 2.54 cm Difficulty: 2

4. scientific notation

   Answer: d) 0.062 yd = 6.2 X 10^{-2} yd Difficulty: 2

5. mega-watts

   Answer: e) 10^6 W Difficulty: 2

Substitute the correct prefix to the fun-word.

f) mill
g) deka
h) peta
i) atto
j) kilo
k) deci
l) pico
m) zept
n) giga
o) tera
p) yott
q) exa
r) peta
s) exa
t) tera

6. 10^{-3} -mouse

   Answer: f) milli Difficulty: 2

7. 10 -cards

   Answer: g) deka Difficulty: 2

8. 10^{15} -pan

   Answer: h) peta Difficulty: 2

Chapter 1

9. 10^{-18} -money

   Answer: i) atto Difficulty: 2

10. 10^3 -mockingbird

    Answer: j) kilo Difficulty: 2

11. 10^{-1} and Lucy

    Answer: k) deci Difficulty: 2

12. 10^{-12} -low

    Answer: l) pico Difficulty: 2

13. 10^{-21}, Harpo, Chico, & Groucho

    Answer: m) zepto Difficulty: 2

14. 10^9 -ly

    Answer: n) giga Difficulty: 2

15. 10^{12} -firma

    Answer: o) tera Difficulty: 2

16. 10^{24} -see me now

    Answer: p) yotta Difficulty: 2

17. $10^{18}2$ -terrestrial

    Answer: q) exa Difficulty: 2

18. 10^{15}, Paul, & Mary

    Answer: r) peta Difficulty: 2

19. 10^{18} -lent

    Answer: s) exa Difficulty: 2

20. 10^{12} -tory

    Answer: t) tera Difficulty: 2

MULTIPLE CHOICE

21. All of the following are base units of the SI system except:
    a) kilogram.
    b) kelvin.
    c) meter.
    d) volt.

    Answer: b Difficulty: 2

22. Which system of units is not metric?
    a) cgs
    b) mks
    c) fps

    Answer: c Difficulty: 2

23. 100 mL is equal to
    a) 1 kL
    b) $10^{-6} \mu L$
    c) 0.1 L
    d) 0.01 ML

    Answer: c Difficulty: 2

24. Using dimensional analysis, which one of the following equations is dimensionally correct?
    ($a \Rightarrow m/s^2$, $v \Rightarrow m/s$, $x \Rightarrow m$)
    a) $v = 2ax$
    b) $x^2 = 2av$
    c) $\sqrt{2x/a}$
    d) $x = v/t$

    Answer: c Difficulty: 2

25. The distance d through which a beam of length L is deflected when it is subjected to load, may be described by the relationship $d = RL^2$. What are the dimensions of the constant R?
    a) meters
    b) $(meters)^{-1}$
    c) $(meters)3$
    d) R is dimensionless.

    Answer: b Difficulty: 2

Chapter 1

26. The ratio R at which paint can be sprayed from a spray gun may be expressed as R = at. If R is measured in m^3/s, and time is measured in seconds, what are the units of "a"?
    a) m^3
    b) $m^3\ s$
    c) m^3/s
    d) m^3/s^2

    Answer: d   Difficulty: 2

TRUE/FALSE

27. Dimensional analysis can tell you whether an equation is physically correct.

    Answer: F   Difficulty: 2

MULTIPLE CHOICE

28. The SI prefix for 10^{-12} is
    a) tera-
    b) giga-
    c) nano-
    d) pico-

    Answer: d   Difficulty: 2

29. Given (1 angstrom unit = 10^{-10} m) and (1 fermi = 10^{-15} m), what is the relationship between these units?
    a) 1 angstrom = 10^5 fermi
    b) 1 angstrom = 10^{-5} fermi
    c) 1 angstrom = 10^{-25} fermi
    d) 1 angstrom = 10^{+25} fermi

    Answer: a   Difficulty: 2

30. All of the following are base units of the SI system except:
    a) kilogram.
    b) kelvin.
    c) meter.
    d) volt.

    Answer: b   Difficulty: 2

ESSAY

31. A cubic box has sides of length 8.0 cm. What is the maximum number of spherical balls of diameter 1.5 cm that can fit inside the closed box ($V_{sphere} = 4/3\ \pi r^3$)?

    Answer: 289   Difficulty: 2

MULTIPLE CHOICE

32. In Einstein's famous equation $E = mc^2$, describing the relationship between matter and energy, the units for E are
    ($m \Longrightarrow$ kg, $c \Longrightarrow$ m/s)
    a) kg m/s
    b) kg m/s^2
    c) m^2/s^2
    d) kg m^2/s^2

    Answer: d Difficulty: 2

33. What is the mass of a woman who weighs 110 lb?
    a) 50 kg
    b) 55 kg
    c) 110 kg
    d) 242 kg

    Answer: b Difficulty: 2

34. mg = kx describes a spring that is stretched by hanging an object on it. The constant k is called the spring constant. What are its units?
    a) kg/s^2
    b) m s^2/kg
    c) kg s^2
    d) s/m kg

    Answer: a Difficulty: 2

35. A football field is 120 yd long and 50 yd wide. What is the area of the football field, in m^2?
    a) 2400 m^2
    b) 3688 m^2
    c) 4206 m^2
    d) 5019 m^2

    Answer: d Difficulty: 2

36. An average human has a heart rate of 70 beats per minute. If someone's heart beats at that average rate over a 70-yr lifetime, how many times would it beat?
    a) 7.44 x 10^5
    b) 2.20 x 10^6
    c) 1.78 x 10^7
    d) 2.58 x 10^9

    Answer: d Difficulty: 2

Chapter 1

TRUE/FALSE

37. One metric ton is 2000 lb.

    Answer: F Difficulty: 2

MULTIPLE CHOICE

38. The last page of a book is numbered 764. The book is 3 cm thick. What is the average thickness of a sheet of paper in the book, in centimeters?
    a) 0.0039
    b) 0.0078
    c) 127.3
    d) 254.7

    Answer: b Difficulty: 2

39. A sunken treasure sits on the ocean floor at a depth of 600 fathoms. What is this depth, in feet? (1 fathom = 6 ft)
    a) 100 ft
    b) 600 ft
    c) 1200 ft
    d) 3600 ft

    Answer: d Difficulty: 2

40. A 400-m tall tower casts a 600-m long shadow over level ground. At what angle θ is the sun elevated above the horizon?
    a) 33.7°
    b) 41.8°
    c) 48.2°
    d) Can't be found; not enough information.

    Answer: a Difficulty: 2

41. An ly (light year) is the distance that light travels in one year. The speed of light is 3.00×10^8 m/s. How many miles are there in an ly? (1 mi = 1609 m), (1 yr = 365 days)
    a) 9.46×10^{12} mi
    b) 9.46×10^{15} mi
    c) 5.88×10^{12} mi
    d) 5.88×10^{15} mi

    Answer: c Difficulty: 2

ESSAY

42. An astronomical unit (Au) is equal to the average distance from the earth to the sun, about 92.9×10^6 mi. A parsec (pc) is the distance at which 1 Au would subtend an angle of 1 second of arc. (a) How many miles are there in a parsec? (b) How many Aus are there in a parsec?

    Answer:
    (a) 1.92×10^{13} miles = 1 parsec
    (b) 2.07×10^5 Aus = 1 parsec
    Difficulty: 2

MULTIPLE CHOICE

43. How many seconds would it take light, traveling at a speed of 186,000 mi/s, to reach us from the sun? The sun is 93,000,000 mi from the earth.
    a) 500 s
    b) 250 s
    c) 1.7×10^{13} s
    d) 2 s

    Answer: a Difficulty: 2

44. Which of the following has three significant figures?
    a) 305.0 cm
    b) 0.0500 mm
    c) 1.00081 kg
    d) 8.060×10^{11} m^2

    Answer: b Difficulty: 2

45. The number of significant figures in 0.40 is
    a) one.
    b) two.
    c) three.
    d) four.

    Answer: b Difficulty: 2

ESSAY

46. Determine the number of significant figures in each of the following measured numbers:
    (a) 4.61 cm   (b) 24.0 s   (c) 0.055 ms   (d) 100.01 m

    Answer: (a) 3   (b) 3   (c) 2   (d) 5 Difficulty: 2

Chapter 1

47. Express each of the following numbers to three significant figures:
(a) 21.22 m (b) 208.7 kg (c) 0.0015601 g (d) 221 s

Answer: (a) 21.2 m (b) 209 kg (c) 0.00156 g (d) 221 s

Difficulty: 2

TRUE/FALSE

48. 0.0097×10^6 cm equals 97×10^{10} cm.

Answer: F Difficulty: 2

MULTIPLE CHOICE

49. 0.00325×10^{-8} cm equals
a) 3.25×10^{-12} mm
b) 3.25×10^{-11} mm
c) 3.25×10^{-10} mm
d) 3.25×10^{-9} mm

Answer: c Difficulty: 2

50. The number of significant figures a common ruler can measure is
a) one.
b) three.
c) five.
d) seven.

Answer: b Difficulty: 2

51. Four students measure the mass of an object, each using a different scale. They record their results as follows:

student	A	B	C	D
mass (g)	49.06	49	50	49.2

Which student used the least precise scale?
a) A
b) B
c) C
d) D

Answer: c Difficulty: 2

ESSAY

52. Express the result of the following calculation, to the proper number of significant figures:
8.37 + 4,240 =

Answer: 4,250 Difficulty: 2

53. Express the result of the following calculation, to the proper number of significant figures:
$50.19 - 7966 \times 10^{-3} =$

Answer: 42.22 Difficulty: 2

54. Express the result of the following calculation, to the proper number of significant figures:
$(0.02739) \times (-240,000) =$

Answer: -6,600 Difficulty: 2

55. Express the result of the following calculation, in scientific notation, to the proper number of significant figures:
$((395600.1)/(6.72)) + 19 =$

Answer: 5.9×10^4 Difficulty: 2

56. Add the following lengths, each obtained from a different measuring instrument, and round off the answer to the proper number of significant figures: 20.02 m, 5.91 m, 0.0097 m, and 2.4669 m.

Answer: 29 m Difficulty: 2

57. A group of students performed an experiment to measure the density of a block of material. They obtained an experimental value of 2.78 g/cm^3. They looked up the standard value for the material in the Handbook of Physics and Chemistry, and found it to be 3.04 g/cm^3. What was their percent error?

Answer: 8.55% Difficulty: 2

58. An experiment designed to determine the volume of a metal statue resulted in an experimental volume 23% greater than its standard value of 1.52 cm^3. What was the experimental value of the volume?

Answer: 1.75 cm^3 Difficulty: 2

59. Evelyn, Marty, and Mary performed a delicate experiment that allowed them to determine how long a cosmic ray particle lived, after it entered the atmosphere from outer space. Their experimental result was 5.22 μs, which was an error of -15.5% compared to the standard accepted value. What is the standard accepted value?

Answer: 6.18 μs Difficulty: 2

Chapter 1

MULTIPLE CHOICE

60. Wall posters are usually sold curled up in cylindrical cardboard tubes. If the length of the tube is 84.5 cm, and the diameter of the tube is 2.40 cm, what is the area of the poster, in cm^2? (Assume the poster doesn't overlap itself.)
    a) 203 cm^2
    b) 382 cm^2
    c) 637 cm^2
    d) 1529 cm^2

    Answer: c Difficulty: 2

61. 0.0001993 is the same as
    a) 1.993×10^{-4}
    b) 19.93×10^5
    c) 1993×10^7
    d) 199.3×10^2

    Answer: a Difficulty: 2

TRUE/FALSE

62. A 2-L bottle of soda gives you more for your money than a 2-qt bottle would, at the same price.

    Answer: T Difficulty: 2

MULTIPLE CHOICE

63. I ran my fastest marathon (42.0 km) in $2^{hrs}:57^{min}$. My average speed, in m/s, was
    a) 14.2×10^3 m/s
    b) 124 m/s
    c) 3.95 m/s
    d) 14.2 m/s

    Answer: c Difficulty: 2

64. The average density of blood is 1.06×10^3 kg/m^3. If you donate a pint of blood to the Red Cross, what mass of blood have you donated, in grams?
    (1 pt = 1/2 L, 1 L = 1000 cm^3)
    a) 530 g
    b) 0.530 g
    c) 5.30×10^3 g
    d) 5.30×10^5 g

    Answer: a Difficulty: 2

ESSAY

65. The mass of Mars, 6.40×10^{23} kg, is about one-tenth that of the earth, and its radius, 3395 km, is about half that of earth. What is the mean density of Mars?

    Answer: 3900 kg/km³   Difficulty: 2

MULTIPLE CHOICE

66. A car travels at 40 km/h for 30 min and 60 km/h for 15 min. How far does it travel in this time?
    a) 20 km
    b) 35 km
    c) 37.5 km
    d) 45 km

    Answer: b Difficulty: 2

67. Concrete is sold by the cubic yard. What is the mass, in kilograms, of one cubic yard of concrete that is five times as dense as water? (1 m = 1.094 yd, and 1 m³ of water has a mass of 1,000 kg.)
    a) 764 kg
    b) 2420 kg
    c) 3819 kg
    d) 6546 kg

    Answer: d Difficulty: 2

68. What are the dimensions of a square acre of ground if 1 square mile = 640 acres?
    a) 147' x 147'
    b) 156' x 156'
    c) 194' x 194'
    d) 209' x 209'

    Answer: d Difficulty: 2

69. A porch roof sloping at 45° accumulated snow to a depth of 30 cm. The roof measured 3 m x 5 m. If the density of snow was 15 kg/m³, what was the weight of the snow, in pounds, on the roof?
    a) 74 lb
    b) 105 lb
    c) 148 lb
    d) 210 lb

    Answer: b Difficulty: 2

Chapter 1

ESSAY

70. A thick-walled metal pipe of length 20 cm has an inside diameter of 2.0 cm and an outside diameter of 2.4 cm. What is the total surface area of the pipe, counting the ends?

    Answer: 279 cm² Difficulty: 2

71. A 2-qt bottle of soda is on sale for $1.29. What should be the price of a 2-Liter bottle of the same soda to yield the same value?

    Answer: $1.36 Difficulty: 2

72. The earth is divided into 24 time zones, in each of which local time differs by 1 hour. At 45° N latitude, what is the width of a time zone? (The earth's radius is 6370 km. 45° N latitude is about as far north as Portland, Oregon, and is halfway between the equator and the north pole.)

    Answer: 1179 km Difficulty: 2

73. Why so much emphasis upon units? Why are units considered to be as important as the quantity, "magnitude", of something?

    Answer:
    Just knowing the magnitude of something is not complete information (in fact it is ambiguous) unless one also knows the units. For example learning that you will earn 150 for a certain task might mean 150 dollars, or perhaps 150 cents, or 150 pesos...... The magnitude is almost worthless without the unit in which it is expressed.
    Difficulty: 2

MULTIPLE CHOICE

74. Which of the following properties of an object does not change when it is moved from the surface of the Earth to the surface of the moon?

    a) age
    b) weight
    c) mass
    d) position

    Answer: c Difficulty: 1

FILL-IN-THE-BLANK

75. The letters c, g, and s in the "cgs system" stand for _____, _____, and _____ ?

    Answer: centimeter, gram, and second Difficulty: 1

MULTIPLE CHOICE

76. Which one of the following volumes is largest?
    a) cc
    b) liter
    c) quart
    d) 500. cm^3

    Answer: b Difficulty: 1

77. Sixty miles/hour is about how many meters/second?
    a) 96
    b) 112
    c) 1.6
    d) 27

    Answer: d Difficulty: 1

78. What is the area of a circle of radius 3.540 meters?
    a) 3.937
    b) 3.94
    c) 3.9369
    d) 3.93692

    Answer: a Difficulty: 1

ESSAY

79. Define and explain the meaning of DENSITY.

    Answer:
    Density expresses how much mass is contained in a unit of volume. i.e.
    Density = Mass / Volume

    Difficulty: 1

MULTIPLE CHOICE

80. How many significant figures are in 0.0037010?
    a) 4
    b) 5
    c) 6
    d) 7
    e) 8

    Answer: b Difficulty: 1

Chapter 1

81. The SI unit of time, the second, is defined by
    a) the motion of the Earth in orbit.
    b) the rotation period of the Earth.
    c) a duration of an atomic clock.
    d) a standard clock at the International Bureau of Weights and Measures in France.

    Answer: c Difficulty: 1

82. Which of the following has the greatest number of significant figures?
    a) 03.1400
    b) 3.142
    c) 0.003142
    d) 314.1
    e) 000.3140

    Answer: a Difficulty: 1

ESSAY

83. What distinguishes a derived unit from a base unit?

    Answer: A derived unit can be expressed in terms of a combination of base uni

    Difficulty: 2

FILL-IN-THE-BLANK

84. If kerosene costs 130. Passi per liter, and one dollar buys 227. Passi; what is its cost in DOLLARS per GALLON? [note 1.00 gallon = 3.79 liters]

    Answer: $ 2.17 per gallon Difficulty: 2

85. When referring to the "mks" system, what does the "m", the "k", and the "s" stand for?

    Answer: meter, kilogram, second Difficulty: 1

86. If we learn that $F = G\, mM/d^2$ where the units are F:Newtons, m:kg, M:kg, and d:meters; what are the units of G?

    Answer: Newton-meter^2 / kg^2 Difficulty: 1

14

Testbank

MULTIPLE CHOICE

87. If we find $v = A\lambda$, where λ is a length and v is a speed, what are the mks units for the A?
    a) s
    b) s^{-1}
    c) m/s^2
    d) m^2/s
    e) kg-m/s

    Answer: b   Difficulty: 1

FILL-IN-THE-BLANK

88. Consider a candy bar which contains 4500. kilocalories. Convert 4500. kcal to Mega electronvolts.   1.000 calorie = 4.186 Joules, 1.000 electronvolt = 1.602×10^{-19} Joules

    Answer: 4500. kcal = 1.176×10^{20} MeV   Difficulty: 2

MULTIPLE CHOICE

89. The METER is currently defined:
    a) by a standard metal bar kept in Paris, France.
    b) as 10^{-7} of the distance from Equator to Pole.
    c) by the President's foot.
    d) by the distance light will travel in a certain time.
    e) by a platinum bar kept in the National Bureau of Standards.

    Answer: d   Difficulty: 1

ESSAY

90. Why is MASS considered a more basic property than WEIGHT?

    Answer:
    A mass may have weight but the weight is a property which changes as the mass is moved around in the universe. Mass is not a property which depends upon position.

    Difficulty: 1

91. Give some examples where metric measures have become common units of every-day measures.

    Answer:
    Examples: Softdrinks are measured in LITERS. Many times periods are expressed in SECONDS. Some road signs give distances in KILOMETERS. Some foodstuffs express their amount in GRAMS. Radio frequencies are expressed in KILOHERTZ. etc.

    Difficulty: 2

Chapter 1

MULTIPLE CHOICE

92. How many cm^2 equals a m^2?
    a) 10^{-4}
    b) 10^{-2}
    c) 10^2
    d) 10^4
    e) 10^6
    f) 10^8

    Answer: d Difficulty: 2

FILL-IN-THE-BLANK

93. Consider the equation $Y = X \ln(Az)$ where Y, X, and z are in meters. What are the units for the constant A?

    Answer: m^{-1} because the argument of a logarithm is dimensionless.

    Difficulty: 2

Chapter 2

MATCHING

Match the physical unit to the the physical quantity.

a) 40 km, SW
b) 9.8 m/s²
c) -120 mi/S
d) 32 ft/s²
e) 186,000 mi

1. displacement

   Answer: a)   40 km, SW   Difficulty: 2

2. SI acceleration of gravity

   Answer: b)   9.8 m/s²   Difficulty: 2

3. velocity

   Answer: c)   -120 mi/S   Difficulty: 2

4. fps acceleration of gravity

   Answer: d)   32 ft/s²   Difficulty: 2

5. distance

   Answer: e)   186,000 mi   Difficulty: 2

MULTIPLE CHOICE

6. Which of the following can never be negative?
   a) Average velocity
   b) Displacement
   c) Instantaneous speed
   d) Acceleration of gravity

   Answer: c   Difficulty: 2

7. If you run a complete loop around an outdoor track (400 m), in 100 s, your average velocity is
   a) 0.25 m/s
   b) 40 m/s
   c) 40,000 m/s
   d) zero

   Answer: d   Difficulty: 2

Chapter 2

   8. All of the following are scalars, except:
   a) mass.
   b) force.
   c) temperature.
   d) distance.

   Answer: b Difficulty: 2

TRUE/FALSE

9. It is possible to have constant speed, but still be accelerating.

   Answer: T Difficulty: 2

10. 60 mi/h equals 88 ft/s.

    Answer: T Difficulty: 2

11. It is possible to have a zero acceleration, and still be moving.

    Answer: T Difficulty: 2

MULTIPLE CHOICE

12. When is the average velocity of an object equal to the instantaneous velocity?
    a) This is always true.
    b) This is never true.
    c) This is the case only when the velocity is constant.
    d) This is the case only when the velocity is increasing at a constant rate.

    Answer: c Difficulty: 2

ESSAY

13. A race car circles 10 times around an 8-km track in 20 min. (a) What is its average speed per lap? (b) What is its average velocity per lap?

    Answer: (a) average lap speed = 66.7 m/s   (b) average lap velocity = 0

    Difficulty: 2

MULTIPLE CHOICE

Figure 2-1

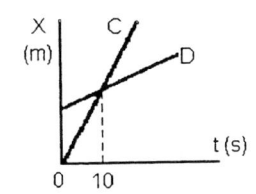

14. Refer to Figure 2-1, At t = 0 s
    a) rider C is ahead of rider D.
    b) rider D is ahead of rider C.
    c) riders C and D are at the same position.

    Answer: b Difficulty: 2

15. Refer to Figure 2-1, At t = 0 s
    a) C is moving, and D is at rest.
    b) D is moving, and C is at rest.
    c) C and D are both moving.
    d) C and D are both at rest.

    Answer: c Difficulty: 2

16. Refer to Figure 2-1, At t = 0 s
    a) C has a greater velocity than D.
    b) D has a greater velocity than C.
    c) C and D have the same velocity.
    d) C is accelerating.

    Answer: a Difficulty: 2

17. Refer to Figure 2-1, At t = 10 s
    a) C and D are at the same position.
    b) C and D have the same velocity.
    c) the velocity of D is greater than the velocity of C.
    d) C is in front of D.

    Answer: a Difficulty: 2

Chapter 2

Figure 2-2

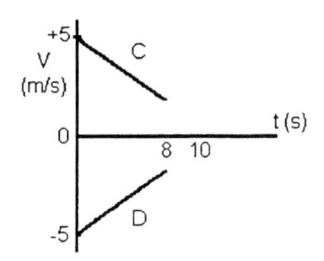

[Graph: V (m/s) vs t(s). Line C starts at +5 and decreases to 0 at t=8. Line D starts at 0 and decreases to -5 at t=8, shown to t=10.]

18. Refer to Figure 2-2, During the first 8 s
    a) C is slowing down, and D is speeding up.
    b) C and D are both slowing down.
    c) C and D have constant velocities.
    d) C has the same average velocity as D.

    Answer: a Difficulty: 2

19. Refer to Figure 2-2, During the first 8 s
    a) C always has a greater acceleration than D.
    b) D always has a greater acceleration than C.
    c) their accelerations are equal in magnitude, but opposite in sign.
    d) their accelerations are equal in magnitude, and equal in sign.

    Answer: c Difficulty: 2

20. Refer to Figure 2-2, Based on all the graphical information
    a) they meet at the same position at t = 8 s.
    b) they will meet at the same position at t = 10 s.
    c) they will never meet at the same position.
    d) not enough information is given to decide if they meet.

    Answer: d Difficulty: 2

TRUE/FALSE

21. Negative velocities, approaching zero velocity, are not negative accelerations.

    Answer: F Difficulty: 2

MULTIPLE CHOICE

22. A new car manufacturer advertises that their car can go "from zero to sixty in 8 s". This is a description of
    a) average speed.
    b) instantaneous speed.
    c) average acceleration.
    d) instantaneous acceleration.

    Answer: c Difficulty: 2

23. Can an object's velocity change direction when its acceleration is constant?
    a) No, this is not possible because it is always speeding up.
    b) No, this is not possible because it is always speeding up or always slowing down, but it can never turn around.
    c) Yes, this is possible, and a rock thrown straight up is an example.
    d) Yes, this is possible, and a car that starts from rest, speeds up, slows to a stop, and then backs up is an example.

    Answer: c Difficulty: 2

24. Can an object have increasing speed while its acceleration is decreasing?
    a) No, this is impossible because of the way in which acceleration is defined.
    b) No, because if acceleration is decreasing the object will be slowing down.
    c) Yes, and an example would be an object falling in the absence of air friction.
    d) Yes, and an example would be an object released from rest in the presence of air friction.

    Answer: d Difficulty: 2

25. Suppose that an object is moving with constant acceleration. Which of the following is an accurate statement concerning its motion?
    a) In equal times its speed increases by equal amounts.
    b) In equal times its velocity changes by equal amounts.
    c) In equal times it moves equal distances.
    d) None of the above is true.

    Answer: c Difficulty: 2

26. A can, after having been given a kick, moves up along a smooth hill of ice. It will
    a) travel at constant velocity.
    b) have a constant acceleration up the hill, but a different constant acceleration when it comes back down the hill.
    c) have the same acceleration, both up the hill and down the hill.
    d) have a varying acceleration along the hill.

    Answer: c Difficulty: 2

TRUE/FALSE

27. When a ball is thrown straight up, its acceleration at the top is zero.

    Answer: F Difficulty: 2

Chapter 2

MULTIPLE CHOICE

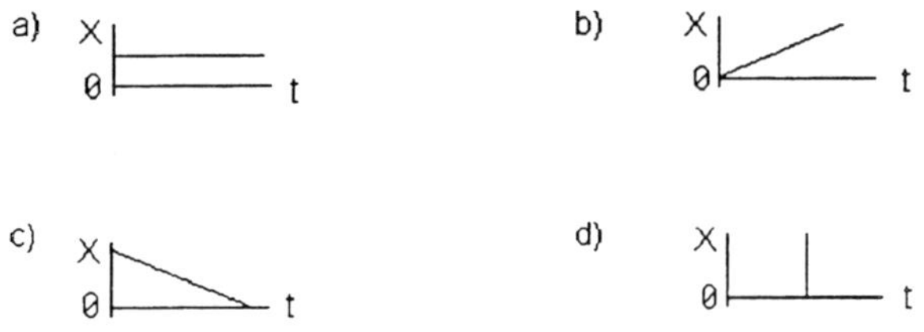

a) [graph: X vs t, horizontal line above 0]
b) [graph: X vs t, line increasing from 0]
c) [graph: X vs t, line decreasing to 0]
d) [graph: X vs t, vertical line]

28. Which graph represents an object at rest?

    Answer: a Difficulty: 2

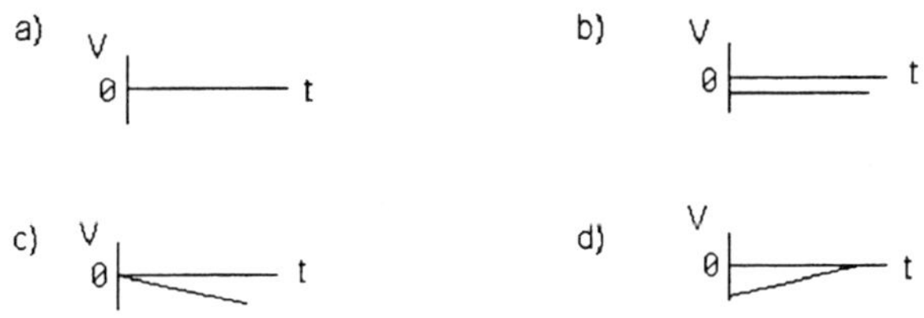

a) [graph: V vs t, horizontal line above 0]
b) [graph: V vs t, horizontal line above 0]
c) [graph: V vs t, line decreasing]
d) [graph: V vs t, line increasing from 0]

29. Which graph represents constant positive acceleration?

    Answer: d Difficulty: 2

30. Under what condition is average velocity equal to the average of the object's initial and final velocity?
    a) The acceleration must be constantly changing.
    b) The acceleration must be constant.
    c) This can only occur if there is no acceleration.

    Answer: b Difficulty: 2

31. When an object is released from rest and falls in the absence of friction, which of the following is true concerning its motion?
    a) Its acceleration is constant.
    b) Its velocity is constant.
    c) Neither its acceleration nor its velocity is constant.
    d) Both its acceleration and its velocity are constant.

    Answer: a Difficulty: 2

32. A skydiver jumps from a high-flying plane. When she reaches terminal velocity, her acceleration
    a) is essentially zero.
    b) is in the upward direction.
    c) is approximately 9.8 m/s^2 downward.

    Answer: a Difficulty: 2

33. Which of the following speeds is greatest?
    a) 10 km/h
    b) 10 mi/h
    c) 10 m/s
    d) 10 ft/s

    Answer: c Difficulty: 2

34. For this problem, assume that the acceleration of gravity is 10 m/s² downward, and that all friction effects can be neglected. A ball is thrown upward at a velocity of 20 m/s. What is its velocity after 3 s?
    a) 10 m/s upward
    b) 10 m/s downward
    c) zero
    d) none of the above choices is correct

    Answer: b Difficulty: 2

35. A motorist travels for 3 h at 80 km/h and 2 h at 100 km/h. What is her average speed for the trip?
    a) 85 km/h
    b) 88 km/h
    c) 90 km/h
    d) 92 km/h

    Answer: b Difficulty: 2

36. A motorist travels 160 km at 80 km/h and 160 km at 100 km/h. What is the average speed of the motorist for this trip?
    a) 84 km/h
    b) 89 km/h
    c) 90 km/h
    d) 91 km/h

    Answer: b Difficulty: 2

37. A boat can move at 30 km/h in still water. How long will it take to move 12 km upstream in a river flowing 6 km/h?
    a) 20 min
    b) 22 min
    c) 24 min
    d) 30 min

    Answer: d Difficulty: 2

Chapter 2

38. Consider a boat that can travel with speed V in still water. For which of the following trips will the elapsed time be least?
   a) Boat travels a distance 2d in still water.
   b) Boat travels a distance d upstream and then returns to its starting point.
   c) Boat travels a distance d downstream and then returns to its starting point.
   d) The time is the same in all of the above cases.

   Answer: a Difficulty: 2

ESSAY

39. In a 400-m relay race the anchorman (the person who runs the last 100 m) for the Trojans can run 100 m in 9.8 s. His rival, the anchorman for the Bruins, can cover 100 m in 10.1 s. What is the largest lead the Bruin runner can have when the Trojan runner starts the final leg of the race, in order that the Trojan runner not lose the race?

   Answer: 3 m Difficulty: 2

40. Suppose that a Ferrari and a Porsche begin a race with a moving start, and each moves with constant speed. One lap of the track is 2 km. The Ferrari laps the Porsche after the Porsche has completed 9 laps. If the speed of the Ferrari had been 10 km/h less, the Porsche would have traveled 18 laps before being overtaken. What were the speeds of the two cars?

   Answer: Porsche speed = 180 km/h; Ferrari speed = 200 km/h

   Difficulty: 2

41. A bat, flying due east at 2 m/s, emits a shriek that is reflected back to it from an oncoming insect flying directly toward the bat at 4 m/s. The insect is 20 m from the bat at the instant the shriek is emitted. Sound travels 340 m/s in air. After what elapsed time does the bat hear the reflected echo?

   Answer: 57 ms (milliseconds) Difficulty: 2

MULTIPLE CHOICE

42. An airplane increases its speed from 100 m/s to 160 m/s, at the average rate of 15 m/s^2. How much time does it take for the complete increase in speed?
   a) 17.3 s
   b) 0.0577 s
   c) 4.0 s
   d) 0.25 s

   Answer: c Difficulty: 2

43. A car traveling 30 mi/h is able to stop in a distance d. Assuming the same braking force, what distance does this car require to stop when it is traveling twice as fast?
   a) d
   b) $\sqrt{2}$ d
   c) 2 d
   d) 4 d

   Answer: d Difficulty: 2

44. A car decelerates uniformly and comes to a stop after 10 s. The car's average velocity during deceleration was 50 km/h. What was the car's deceleration while slowing down?
   a) 10 km/h-s
   b) 8 km/h-s
   c) 5 km/h-s
   d) 4 km/h-s

   Answer: a Difficulty: 2

ESSAY

45. A car traveling 60 km/h accelerates at the rate of 2 m/s². How much time is required for the car to reach a speed of 90 km/h?

   Answer: 4.17 s Difficulty: 2

46. A bullet moving horizontally with a speed of 500 m/s strikes a sandbag and penetrates a distance of 10 cm.
   (a) What is the average acceleration of the bullet?
   (b) How long does it take to come to rest?

   Answer:
   (a) 1.25 x 10⁶ m/s²
   (b) 0.4 ms
   Difficulty: 2

MULTIPLE CHOICE

47. In which of the following cases will a car move the greatest distance?
   a) A car with speed v_1 travels with constant speed for time t_1 and then accelerates uniformly for t_2 s to speed v_2.
   b) A car with speed v_1 accelerates uniformly to speed v_2 in t_1 seconds and then travels at speed v_2 for t_2 seconds.
   c) It is not possible to answer this question definitively without knowing numerical values for t_1, t_2, v_1, and v_2.

   Answer: b Difficulty: 2

Chapter 2

ESSAY

48. A car with good tires on a dry road can decelerate at about 5 m/s² when braking. If the car is traveling at 55 mi/h, (a) how long does it take the car to stop under these conditions? (b) how far does the car travel during this time?

    Answer: (a) 4.9 s   (b) 60.4 m   Difficulty: 2

49. In a test carried out by a car manufacturer, a test driver is asked to put on his brakes when a warning light is suddenly flashed on in the roadway ahead. When traveling 25 m/s the driver is able to stop in 98 m, and when traveling 10 m/s he is able to stop in 32.5 m. Assuming the driver's reaction time is the same in each case, and that the rate of deceleration when the brakes are applied is independent of speed, determine (a) the driver's reaction time. (b) the driver's deceleration.

    Answer: (a) 2.8 s   (b) 11.2 m/s²   Difficulty: 2

MULTIPLE CHOICE

50. A jet fighter plane is launched from a catapult on an aircraft carrier. It reaches a speed of 42 m/s at the end of the catapult, and this requires 2 s. Assuming the acceleration is constant, what is the length of the catapult?
    a) 16 m
    b) 24 m
    c) 42 m
    d) 84 m

    Answer: c Difficulty: 2

51. A car starting from rest moves with constant acceleration of 2 m/s² for 10 s, then travels with constant speed for another 10 s, and then finally slows to a stop with constant acceleration of -2 m/s². How far does it travel?
    a) 200 m
    b) 300 m
    c) 400 m
    d) 500 m

    Answer: c Difficulty: 2

52. A car starts from rest and accelerates uniformly at 3 m/s². A second car starts from rest 6 s later at the same point and accelerates uniformly at 5 m/s². How long does it take the second car to overtake the first car?
a) 12.2 s
b) 18.9 s
c) 22.7 s
d) 24.0 s

Answer: c Difficulty: 2

53. A toy rocket is launched upward with a net acceleration of 10 m/s² for 3 s. It then slows at the rate of 10 m/s² until it reaches its maximum altitude. How high does it go?
a) 30 m
b) 45 m
c) 60 m
d) 90 m

Answer: c Difficulty: 2

54. A bullet shot straight up returns to its starting point in 10 s. Its initial speed was
a) 9.8 m/s
b) 24.5 m/s
c) 49 m/s
d) 98 m/s

Answer: c Difficulty: 2

ESSAY

55. A ball is thrown straight up with a speed of 36 m/s. How long does it take to return to its starting point?

Answer: 2.71 s Difficulty: 2

56. Contrast Aristotle's predictions concerning free-falling bodies with Galileo's predictions.

Answer:
Aristotle said heavier bodies fall faster than light bodies; so if the two were dropped at the same time, the heavier body would strike the ground sooner. Galileo observed that the motion was independent of the body's mass provided air friction was not significant.

Difficulty: 2

Chapter 2

FILL-IN-THE-BLANK

57. An object moving under the influence of only gravity is said to be in _____ _____.

    Answer: free fall   Difficulty: 2

58. "Big Mike" throws a baseball straight up and it eventually falls back to him. When the ball was at its highest point, what was its velocity and what was its acceleration?

    Velocity=_____

    Acceleration=_____

    (remember to include magnitude and direction)

    Answer: velocity=zero ; acceleration was 9.8 m/s^2 downward

    Difficulty: 2

59. Captain Rickard orders his starship to accelerate from rest at "1g" (accel=9.8 m/s^2). How long does it take the starship to reach one-tenth the speed of light ? [light travels 300. megameters/s]

    Answer: 35.days   Difficulty: 2

60. An astronaut on a strange new planet finds that she can jump up to a maximum height of 27. meters when her initial upward speed is 6.0 m/s. What is the magnitude of the acceleration of gravity on the planet?

    Answer: 0.67 m/s^2 Difficulty: 2

61. A bullet moving at 244.m/s strikes a tree and penetrates a distance of 8.34 mm before stopping.

    a) What was the average acceleration of the bullet as it slowed?
    b) Assuming a constant acceleration, how long did it take the bullet to stop?

    Answer: a) -3.57 x 10^6 m/s^2 b) 68.4 microseconds   Difficulty: 2

ESSAY

62. In the formula $v^2 = v_1^2 + 2 a x$ what does x represent?

    Answer: The CHANGE in POSITION, which is NOT necessarily the distance gone.

    Difficulty: 1

MULTIPLE CHOICE

63. If an object is accelerating, it must therefore undergo:
   a) a change in direction.
   b) a change in velocity.
   c) a decrease in velocity.
   d) an increase in velocity.

   Answer: b Difficulty: 1

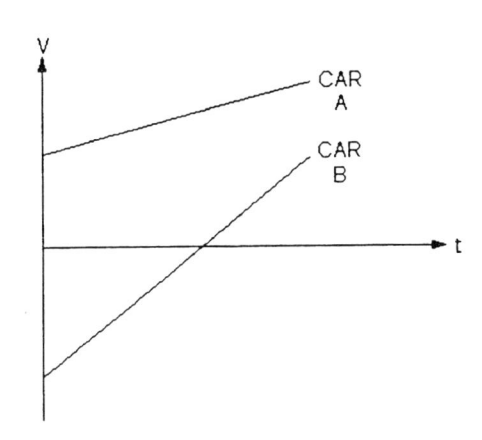

64. The accompanying graph plots the velocity of two cars (A & B) along the same straight road. During the time interval shown, which car is AHEAD?
   a) A
   b) B
   c) insufficient information

   Answer: c Difficulty: 2

SHORT ANSWER

65. Eric watches a jet powered truck during an "air-show". It accelerates from rest to 300. mph in 6.0 seconds. The accelerating was equivalent to how many "g's"?

   Answer: 22. m/s^2 = 2.3 g's Difficulty: 2

MULTIPLE CHOICE

66. Barbara travels 20.km Northward, then travels 40.km Eastward, then continues for 50.km in a Southward direction. What displacement will now take her back to her initial position?
   a) 30.km Westward + 40.km Northward
   b) 50.km Northwest
   c) 40.km Westward + 30.km Northward
   d) 30.km Northward + 30.km Westward
   e) 40.km Eastward + 30.km Southward
   f) 30.km Eastward + 40.km Southward

   Answer: c Difficulty: 2

Chapter 2

FILL-IN-THE-BLANK

67. Denis swims the length of a 40.meter pool in 10.seconds and immediately swims back to the starting position in another 15.seconds. a) What was his average speed?
b) What was his average velocity?

Answer: a) avg speed 3.2 m/s   b) avg velocity 0   Difficulty: 2

ESSAY

68. Discuss whether a car speedometer measures instantaneous speed or does it determine an average speed?

Answer:
One desires it to display instantaneous speed, but it actually records an average speed. It is designed so that the average is over a very short time so that it closely approximates the instantaneous speed.
Difficulty: 2

69. Explain how a POSITIVE acceleration could produce a DECELERATION.

Answer:
An object which was moving in the negative direction would slow down if its acceleration were positive (a negative acceleration would mean the velocity was getting more negative: it would be speeding up in the negative direction).
Difficulty: 2

70. Consider a heavy object which is thrown straight up, reaches its highest point, and then falls back down to the ground. During what parts of the trajectory was it in "FREE FALL"? (assume here that air friction is negligible)

Answer:
During its entire travel during which gravity was the only significant influence on it (it "freely fell" moving up, moving down, and at its motionless highest point).
Difficulty: 2

Testbank

FILL-IN-THE-BLANK

71. Give examples of four quantities which are:
    a) scalars
    b) vectors

    Answer:
    a) distance, temperature, speed, age
    b) displacement, velocity, acceleration, force
    Difficulty: 2

72. A train takes 3.0 hours to travel the 1st leg of a trip at 20. m/s. How fast must it move over the 2nd leg of 200. km to have a total average speed of 30. m/s ?

    Answer: 65. m/s   Difficulty: 2

73. The Earth-Moon distance is about 238,000. miles. If laser measurements were to show that the Moon is moving away from the Earth at 9.5 cm per year, this would infer that the Moon was very close to the Earth how long ago (assume a constant rate of recession)?

    Answer: 4.0 billion yrs (4.0×10^9 yr)   Difficulty: 2

MULTIPLE CHOICE

74. Who said: "I think that in the discussion of natural problems we ought not to begin at the authority of places of Scripture, but at sensible experiments ...."?
    a) Einstein
    b) Aristotle
    c) Viviani
    d) Galileo

    Answer: d   Difficulty: 2

TRUE/FALSE

75. The "acceleration of gravity" is 9.80 m/s^2 everywhere on the surface of the Earth.

    Answer: F   Difficulty: 1

Chapter 2

MULTIPLE CHOICE

76. A ball is released moving straight up at 15.m/s. At some time later it is falling downward at 15.m/s. What was the magnitude of its average velocity over this time period ?
    a) 0. m/s
    b) 7.5 m/s
    c) 15. m/s
    d) 30. m/s

    Answer: a Difficulty: 2

77. A ball is released moving straight up at 15.m/s. A short time later it is falling downward at 15.m/s. What was the magnitude of its AVERAGE SPEED over this time period ?
    a) 0. m/s
    b) 7.5 m/s
    c) 15. m/s
    d) 30. m/s

    Answer: b Difficulty: 2

78. If a boat can move at 34.mph in still water, how many minutes will it take for it to travel 13. miles upstream in a river flowing at 8.0 mph?
    a) 22.
    b) 24.
    c) 28.
    d) 30.
    e) 32.

    Answer: d Difficulty: 2

Chapter 3

MATCHING

Match the description to the physical quantity.

a) $(\mathbf{v}_o + \mathbf{v})/2$
b) $\mathbf{v}_o + \mathbf{a}\, t$
c) $\mathbf{v}_{AB} + \mathbf{v}_{BC}$
d) $(\mathbf{v} - \mathbf{v}_o)/t$
e) $\mathbf{x}$
f) R
g) parabola
h) $\mathbf{R} + \mathbf{S} + \mathbf{T}$

1. average velocity

    Answer: a) $(\mathbf{v}_o + \mathbf{v})/2$ Difficulty: 2

2. instantaneous velocity

    Answer: b) $\mathbf{v}_o + \mathbf{a}\, t$ Difficulty: 2

3. relative velocity

    Answer: c) $\mathbf{v}_{AB} + \mathbf{v}_{BC}$ Difficulty: 2

4. average acceleration

    Answer: d) $(\mathbf{v} - \mathbf{v}_o)/t$ Difficulty: 2

5. unit-vector

    Answer: e) $\mathbf{x}$ Difficulty: 2

6. range

    Answer: f) R Difficulty: 2

7. trajectory

    Answer: g) parabola Difficulty: 2

8. resultant

    Answer: h) $\mathbf{R} + \mathbf{S} + \mathbf{T}$ Difficulty: 2

Chapter 3

MULTIPLE CHOICE

9. Which of the following is an accurate statement?
   a) A vector cannot have zero magnitude if one of its components is not zero.
   b) The magnitude of a vector can be less than the magnitude of one of its components.
   c) If the magnitude of vector A is less than the magnitude of vector B, then the x-component of A is less than the x-component of B.
   d) The magnitude of a vector can be positive or negative.

   Answer: a Difficulty: 2

10. Which of the following operations will not change a vector?
    a) Translate it parallel to itself.
    b) Rotate it.
    c) Multiply it by a constant factor.
    d) Add a constant vector to it.

    Answer: a Difficulty: 2

11. If you drive west at 20 mi/h, then drive east at 15 mi/h, your average velocity will be
    a) 5 mi/h east.
    b) 35 mi/h west.
    c) 35 mi/h east.
    d) 5 mi/h west.
    e) cannot be determined; insufficient information.

    Answer: e Difficulty: 2

12. If the acceleration vector of an object is directed anti-parallel to the velocity vector,
    a) the object is turning.
    b) the object is speeding up.
    c) the object is slowing down.
    d) the object is moving in the negative x-direction.

    Answer: c Difficulty: 2

13. Two displacement vectors have magnitudes of 5 m and 7 m, respectively. When these two vectors are added, the magnitude of the sum
    a) is 12 m.
    b) could be as small as 2 m, or as large as 12 m.
    c) is 2 m.
    d) is larger than 12 m.

    Answer: b Difficulty: 2

14. If the acceleration of an object is always directed perpendicular to its velocity,
    a) the object is speeding up.
    b) the object is slowing down.
    c) the object is turning.
    d) this situation would not be physically possible.

    Answer: c Difficulty: 2

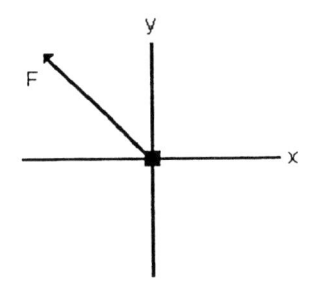

15. 200-lb force is pulling on an object, as shown. The sign of the **x** and **y** components of the force are
    a) **x** (positive), **y** (positive).
    b) **x** (positive), **y** (negative).
    c) **x** (negative), **y** (positive).
    d) **x** (negative), **y** (negative).

    Answer: c Difficulty: 2

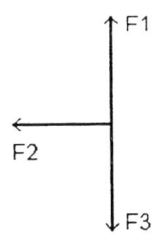

16. Three boys each pull with a 20-lb force on the same object. The resultant force will be
    a) zero.
    b) 20 lb to the left.
    c) 20 lb up.
    d) 20 lb down.

    Answer: b Difficulty: 2

17. If you want to cross a river in a motorboat in the least amount of time, you should head
    a) straight across the river.
    b) slightly upstream.
    c) slightly downstream.
    d) the time will be independent of how you head your boat.

    Answer: b Difficulty: 2

Chapter 3

18. On a calm day (no wind), you can run a 1500-m race at a velocity of 4.0 m/s. If you ran the same race on a day when you had a constant tailwind of 2.0 m/s, the time it would take you to finish would be
    a) 250 s
    b) 750 s
    c) 1125 s
    d) 9000 s

    Answer: a Difficulty: 2

19. Consider a plane flying with groundspeed V_g and airspeed V_a in a wind with velocity V. Which of the following relationships is true?
    a) $V_g = V_a + V$
    b) V_g can be greater than $V_a + V$
    c) V_g can be less than $V_a + V$
    d) V_g can have any value between $V_a + V$ and $V_a - V$

    Answer: d Difficulty: 2

ESSAY

20. An airplane with an airspeed of 120 km/h is headed 30° east of north in a wind blowing due west at 30 km/h. What is the groundspeed of the plane?

    Answer: 102 km/h Difficulty: 2

MULTIPLE CHOICE

21. A fighter plane moving 200 m/s fires a projectile with speed 50 m/s in a forward direction. The projectile makes an angle of 30° with the direction of the plane's motion. What is the speed of the projectile with respect to a stationary observer on the ground?
    a) 245 m/s
    b) 250 m/s
    c) 268 m/s
    d) 293 m/s

    Answer: a Difficulty: 2

22. A swimmer heading directly across a river 200 m wide reaches the opposite bank in 6 min 40 s. She is swept downstream 300 m. How fast can she swim in still water?
    a) 0.50 m/s
    b) 1.24 m/s
    c) 1.42 m/s
    d) 1.83 m/s

    Answer: a Difficulty: 2

ESSAY

23. The driver of a motorboat that can move at 10 m/s wishes to travel directly across a narrow strait in which the current flows at 5 m/s. (a) At what angle upstream should the driver head the boat? (b) How long will it take to cross a distance of 1.6 km?

    Answer: (a) 26.6°   (b) 179 s Difficulty: 2

MULTIPLE CHOICE

24. A butterfly moves with a speed of v = 12.0 m/s. The x-component of its velocity is 8.00 m/s. The angle between the direction of its motion and the x-axis must be
    a) 30.0°
    b) 41.8°
    c) 48.2°
    d) 53.0°

    Answer: c Difficulty: 2

25. An electron, initially moving at 2.0×10^4 m/s along the +x-axis, experiences a constant acceleration of 10^{10} m/s^2 in the +y direction. How far from its starting point is the electron after 4.0×10^{-6} seconds?
    a) 8.0 m
    b) 11.3 m
    c) 8.0 cm
    d) 11.3 cm

    Answer: d Difficulty: 2

26. Three forces, each having a magnitude of 30 lb, pull on an object in directions that are 120° apart from each other. Which one of the following statements must be true?
    a) The resultant force is zero.
    b) The resultant force is greater than 30 lb.
    c) The resultant force is equal to 30 lb.
    d) The resultant force is less than 30 lb.

    Answer: a Difficulty: 2

Chapter 3

Two vectors, R and S, are known:

R ↓ and S →

If vector S is subtracted from vector R, then the vector T = R - S is

a) ↘   b) ↙   c) ↗   d) ↖

27.

Answer: b Difficulty: 2

28. Your motorboat can move at 30 km/h in still water. How much time will it take you to move 12 km upstream, in a river flowing at 6 km/h?
a) 20 min
b) 22 min
c) 24 min
d) 30 min

Answer: d Difficulty: 2

ESSAY

29. Two vectors **A** and **B** have components (0, 1, 2) and (-1, 3, 1), respectively. (a) Determine the components of the sum of these two vectors. (b) Determine the magnitude of the sum of these two vectors.

Answer: (a) (-1, 4, 3) (b) 5.1 Difficulty: 2

MULTIPLE CHOICE

30. Three vectors, expressed in Cartesian coordinates, are

	x-comp	y-comp
S	-3.50	+4.50
T	0	-6.50
U	+5.50	-2.50

The magnitude fo the resultant vector is
a) 4.92
b) 24.25
c) 16.2
d) 17.5

Answer: a Difficulty: 2

31. If vector **A** = -3.0 **x** - 4.0 **y**, and vector **B** = +3.0 **x** -8.0 **y**, then the magnitude of vector **C** = **A** - **B** is
    a) 13.4
    b) 16
    c) 144
    d) 7.2

    Answer: d Difficulty: 2

ESSAY

32. A particle initially moving with a velocity 2 m/s experiences a constant acceleration of 1 m/s^2 in the x-direction and -2 m/s^2 in the y-direction. What are the velocity components of the particle after 4 s?

    Answer: V_x = 6 m/s; V_y = 8 m/s Difficulty: 2

MULTIPLE CHOICE

33. Bullet G is dropped, from rest, at the same time that another bullet H is fired horizontally from a rifle. If both bullets leave from the same height above the ground, then
    a) bullet G hits the ground before bullet H does.
    b) both bullets hit the ground at the same time.
    c) bullet H hits the ground before bullet G does.
    d) not enough information is given to compare when they hit.

    Answer: b Difficulty: 2

34. A girl throws a rock horizontally, with a velocity of 10 m/s, from a bridge. It falls 20 m to the water below. How far does the rock travel horizontally before striking the water?(Use g = 10 m/s^2)
    a) 14 m
    b) 16 m
    c) 20 m
    d) 24 m

    Answer: c Difficulty: 2

35. Suppose that several projectiles are fired upward. Which one will be in the air longest?
    a) The one with the farthest range, R
    b) The one with the highest maximum elevation, h
    c) The one for which the product of (R)(t)(h) is maximum
    d) The one with the greatest initial velocity

    Answer: b Difficulty: 2

Chapter 3

36. An arrow is shot, from a bow, at an original speed of v_o. When it returns to the same horizontal level, its speed will be
   a) $1/2\ v_o$
   b) v_o
   c) $2\ v_o$
   d) $9.8\ v_o$

   Answer: b   Difficulty: 2

37. A golf ball is hit with an initial velocity of 60 m/s at an angle of 130° above the horizontal. How far does it travel?
   a) 152 m
   b) 160 m
   c) 184 m
   d) 318 m

   Answer: d   Difficulty: 2

38. A package of supplies is dropped from a plane, and one second later a second package is dropped. Neglecting air resistance, the distance between the falling packages will
   a) be constant.
   b) decrease.
   c) increase.
   d) depend on their weight.

   Answer: c   Difficulty: 2

39. A plane flying horizontally at a speed of 50 m/s and at an elevation of 160 m drops a package of supplies. Two seconds later it drops a second package. How far apart will the two packages land on the ground?
   a) 100 m
   b) 162 m
   c) 177 m
   d) 283 m

   Answer: a   Difficulty: 2

40. At what angle should a water-gun be aimed in order for the water to land with the greatest horizontal range?
   a) 0°
   b) 30°
   c) 45°
   d) 90°

   Answer: c   Difficulty: 2

41. The acceleration of gravity on the moon is only one-sixth of that on earth. If you hit a baseball on the moon with the same effort (and at the speed and angle) that you would on earth, the ball would land
   a) the same distance away.
   b) one-sixth as far.
   c) 6 times as far.
   d) 36 times as far.

   Answer: c Difficulty: 2

42. A girl throws a stone straight down from a bridge. The stone leaves her hand with velocity of 8.00 m/s at a height of 12 m above the water below. How much time does it take for the stone to hit the water?
   a) 0.41 s
   b) 0.95 s
   c) 2.58 s
   d) 17.3 s

   Answer: b Difficulty: 2

TRUE/FALSE

43. You toss a ball to your friend. When the ball reaches its maximum altitude, its velocity is zero.

   Answer: F Difficulty: 2

44. The ball you tossed reaches its maximum altitude in exactly half of the time it is in the air.

   Answer: T Difficulty: 2

45. When the ball you tossed reaches its maximum altitude, its acceleration is zero.

   Answer: F Difficulty: 2

MULTIPLE CHOICE

46. You hit a tennis ball over the net. When the ball reaches its maximum height, its speed is
   a) zero.
   b) less than its initial speed.
   c) equal to its initial speed.
   d) greater than its initial speed.

   Answer: b Difficulty: 2

Chapter 3

47. You are serving the volleyball for the second time in a volleyball game. If the ball leaves your hand with twice the velocity it had on your first serve, its horizontal range R (compared to your first serve) would be
   a) $\sqrt{2}$ times as much.
   b) half as much.
   c) twice as much.
   d) four times as much.

   Answer: d Difficulty: 2

48. A pilot drops a bomb from a plane flying horizontally. When the bomb hits the ground, the horizontal location of the plane will
   a) be behind the bomb.
   b) be over the bomb.
   c) be in front of the bomb.
   d) depend on the speed of the plane when the bomb was released.

   Answer: b Difficulty: 2

49. A basketball is thrown with a velocity of 20 m/s at an angle of 60° above the horizontal. What is the horizontal component of its instantaneous velocity at the exact top of its trajectory? (Neglect air friction.)
   a) 10 m/s
   b) 17.3 m/s
   c) 20 m/s
   d) zero

   Answer: a Difficulty: 2

50. A soccerball is kicked with a velocity of 25 m/s at an angle of 45° above the horizontal. What is the vertical component of its acceleration as it travels along its trajectory? (Neglect air friction.)
   a) zero
   b) g downward
   c) g sin (45°) upward
   d) g upward
   e) g cos (45°) downward

   Answer: b Difficulty: 2

51. A stone is thrown horizontally from the top of a tower at the same instant that a ball is dropped vertically. Which object is traveling faster when it hits the level ground below?
   a) It is impossible to tell from the information given.
   b) The stone
   c) The ball
   d) Neither, since both are traveling at the same speed.

   Answer: b Difficulty: 2

52. You hit a handball too hard, and it lands horizontally on top of a roof. If its original velocity was v_0 at an angle θ above the horizontal, then the velocity it lands on the roof with is
    a) zero
    b) $v_0 \sin \theta$
    c) $v_0 \cos \theta$
    d) $2v_0$

    Answer: c   Difficulty: 2

53. A ball thrown horizontally from a point 24 m above the ground, strikes the ground after traveling horizontally a distance of 18 m. With what speed was it thrown? (In this problem use $g = 10$ m/s^2.)
    a) 6.10 m/s
    b) 7.40 m/s
    c) 8.22 m/s
    d) 8.96 m/s

    Answer: c   Difficulty: 2

ESSAY

54. A rifle bullet is fired at an angle of 30° below the horizontal with an initial velocity of 800 m/s from the top of a cliff 80 m high. How far from the base of the cliff does it strike the level ground below?

    Answer: 138 m   Difficulty: 2

55. A child drops a toy at rest from a point 4 m above the ground at the same instant her friend throws a ball upward at 6 m/s from a point 1 m above the ground. At what distance above the ground do the ball and the toy cross paths?

    Answer: 2.78 m   Difficulty: 2

56. A mortar shell is launched with a velocity of 100 m/s at an angle of 30° above horizontal from a point on a cliff 50 m above a level plain below. How far from the base of the cliff does the mortar shell strike the ground?

    Answer: 963 m   Difficulty: 2

57. In attempting to jump up a waterfall, a salmon leaves the water 2 m from the base of the waterfall. With what minimum speed must it leave the water in order just to make it up a waterfall 1.6 m high?

    Answer: 8.96 m/s   Difficulty: 2

Chapter 3

MULTIPLE CHOICE

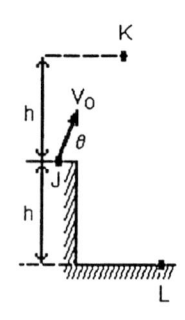

58. You throw a pebble upward as shown, with a velocity v_0 at an angle θ, from a roof (at point J) h ft above he ground. If the pebble rises upward to h ft above the roof (to point K), then falls to the ground below (to point L), its speed when it hits the ground is
    a) v_0
    b) between v_0 and $2v_0$
    c) $2v_0$
    d) greater than $2v_0$

    Answer: c Difficulty: 2

59. An Olympic athlete throws a javelin at four different angles above the horizontal, each with the same speed: 30° 40° 60° 80°. Which two throws cause the javelin to land the same distance away?
    a) 30° and 80°
    b) 40° and 60°
    c) 40° and 80°
    d) 30° and 60°

    Answer: d Difficulty: 2

60. An Air Force plane doing target practice on a floating target sights the target at an angle of 53° from the vertical. The pilot should aim his guns at an angle, from the vertical, of
    a) 53°.
    b) less than 53°.
    c) greater than 53°.
    d) the angle depends on the velocity of the plane.

    Answer: c Difficulty: 2

61. Which of the following is an accurate statement?
    a) Three vectors can never add to zero.
    b) If three vectors add to zero, they must be co-linear (i.e., lie in a straight line).
    c) If three vectors add to zero, they must be in a plane.
    d) If three vectors add to zero, they cannot all have the same magnitude.

    Answer: c Difficulty: 2

44

62. Johnny throws a shot put at an angle of 45.degrees and it travels a horizontal distance of 30.meters. If he releases the shot-put at the same speed but changes the angle to 55.degrees, what will be the new range (distance)?

   a) 25.m
   b) 28.m
   c) 32.m
   d) 37.m
   e) 41.m

   Answer: b Difficulty: 2

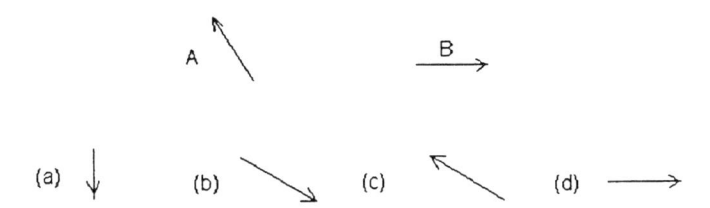

63. Consider two vectors A and B
    The sum of the two vectors is

    Answer: d Difficulty: 2

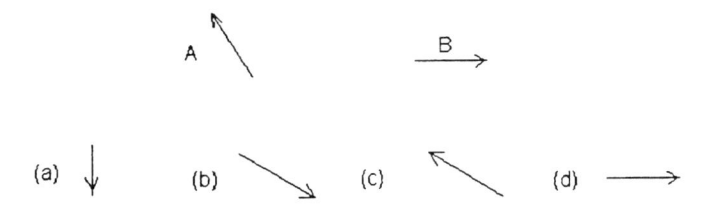

64. Consider two vectors A and B
    The difference A - B is

    Answer: c Difficulty: 2

FILL-IN-THE-BLANK

65. A car and a truck are traveling in the same direction along a straight section of highway. The car is moving at 55.mph and the truck is moving 45.mph.

    a) what is the velocity of the car relative to the truck? _____

    b) what is the velocity of the truck relative to the car? _____

    Answer: a) 10.mph forward,   b) 10.mph backward. Difficulty: 2

Chapter 3

MULTIPLE CHOICE

66. If I walk 8. meters in a straight line and then walk 5. meters in another straight line, the total displacement cannot have a magnitude of:
a) 13.m
b) 8.m
c) 5.m
d) 3.m
e) 2.m

Answer: e Difficulty: 2

67. Consider a thrown ball which has reached its highest point in flight. The magnitude of the acceleration:
a) is zero.
b) is less than it will be a short instant later.
c) reverses direction.
d) is g.

Answer: d Difficulty: 2

FILL-IN-THE-BLANK

68. Lisa throws a stone horizontally from the roof edge of a 50.meter high dormitory. It hits the ground at a point 60.m from the building.
a) Find the time of flight.
b) What was the initial velocity?
c) Determine the horizontal component of velocity just before hitting the ground.
d) What vertical speed did it have just before hitting the ground?

Answer: a) 3.2 s   b) 19.m/s horizontal   c) 19.m/s   d) 31.m/s

Difficulty: 2

69. A hunter points a rifle horizontally 3.3 m above the ground. The bullet leaves the barrel at 325.m/s.
a) How long did it take for the bullet to strike the ground?
b) How far horizontally did it travel?
c) If instead, another bullet is dropped from rest 3.3 m above the ground, how long will it take to fall to the ground?

Answer: a) 0.821 s   b) 267.m   c) 0.821 s Difficulty: 2

70. Holly intends to swim to a point straight across a 100.m wide river that flows at 1.2 m/s.
    a) If she can swim 2.5 m/s in still water, at what angle must she swim upstream to achieve her goal?
    b) How long does it take her to swim to the other side?

    Answer: a) She must angle 29.° upstream.   b) 46.s Difficulty: 2

71. Neglecting the effect of air friction, at what angle should a cannon be raised to achieve the greatest RANGE (horizontal distance)? _____

    Answer: 45. degrees Difficulty: 2

72. A girl throws a rock horizontally with a speed of 12.0 m/s from a bridge. It falls 2.28 s before reaching the water below.
    A) How high is the bridge from the water below?
    B) How far horizontally does the rock travel before striking the water?
    C) When the rock hits the water it is traveling with what horizontal component of velocity?
    D) What was the total speed just before hitting the water?

    Answer: A) 25.0 m/s   B) 27.4 m   C) 12.0 m/s   D) 25.4 m/s

    Difficulty: 2

MULTIPLE CHOICE

73. John can hit a golf ball 310.m on Earth. On the moon's surface $g = 1/6\, g_{earth}$ so how far could he hit the same ball if he could only achieve half the initial velocity achieved on Earth?
    a) 155.m
    b) 52.m
    c) 78.m
    d) 396.m
    e) 465.m
    f) 1860.m

    Answer: e Difficulty: 2

74. When air resistance is a factor, maximum range is achieved when the projection angle is:
    a) less than 45.°
    b) 45.°
    c) greater than 45.°

    Answer: a Difficulty: 2

Chapter 3

FILL-IN-THE-BLANK

75. Consider a velocity with components 36.m/s westward and 22.m/s northward. What is its magnitude and direction?

    Answer: v= 42.m/s at 31.° north of west Difficulty: 2

76. A pickup truck moves 25.m/s eastward. Jerry is standing in the back and throws a baseball in what to him is the southwest direction at 28.m/s (with respect to the truck). A person at rest would see the ball moving HOW FAST in WHAT DIRECTION?

    Answer: 20.m/s   15.° East of South Difficulty: 2

77. A marble moving 1.48 m/s rolls off the top edge of a 125.cm high table.
    a) How far from the base of the table did it strike the floor? b) How long was it "in flight"?

    Answer: a) 75.0 cm   b) 0.505 s Difficulty: 2

MULTIPLE CHOICE

78. A motocycle stunt rider wants to jump a 179.m wide canyon. He arranges a ramp on one side so he will be launched pointed 30 degrees above the horizontal. What minimum speed will be needed to just reach the other side?

    a) 25.m/s
    b) 30.m/s
    c) 35.m/s
    d) 40.m/s
    e) 45.m/s

    Answer: e Difficulty: 2

FILL-IN-THE-BLANK

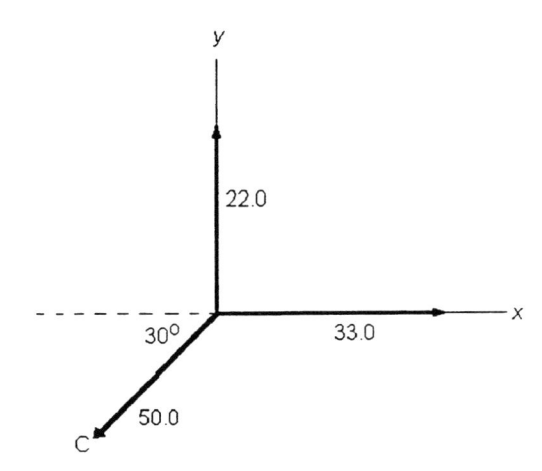

79. Given the three vectors in the figure, determine the vector sum **A + B + C**

    Answer: **A + B + C** = -10**i** -3**j** = 11. at 196.° (third quadrant)

    Difficulty: 2

80. For the three vectors given in problem 3-17, determine the vector "difference" **A - B - C**

    Answer: 76.**i** + 3.**j** = 76. @ 2.3° (1st quadrant)   Difficulty: 2

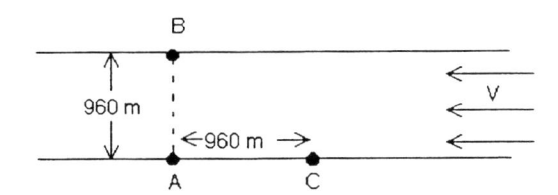

81. A 960.m wide river flows at 16. m/s as shown in the figure. Alice and John have a race in identical boats which travel 20.m/s in still water. Alice leaves point A and steers so that she goes straight to point B directly across and then back to A. John speeds up to point C (960.m upstream) and then returns to A.
    a) which person arrived back at A first?
    b) how much sooner did the winner arrive back at A?

    Answer: a) Alice wins   b) she arrives 107.s before John

    Difficulty: 3

49

Chapter 3

82. An airplane is pointed in the NE direction and is moving 71.m/s with respect to the air. The jetstream (moving air) carries the plane 30.m/s toward the East. What is the velocity of the plane relative to the ground?

    Answer: 94.m/s at 32.° North of East. Difficulty: 2

83. Suppose Chipper Jones hits a home run, for the Braves, which travels 361.feet (110.m). Assuming it left the bat at the optimum 45.°
    a) how fast was it hit?
    b) how high did it go?

    Answer: a) 32.8 m/s = 73.5 mph   b) 27.5 meters = 90.2 feet

    Difficulty: 2

Chapter 4

MATCHING

Match the unit or expression to the physical quantity.

a) dyne
b) N/m
c) equilibrium
d) newton
e) dimensionless
f) slug
g) mass
h) pound
i) cm/s^2

1. cgs unit of force

   Answer: a) dyne Difficulty: 2

2. SI unit of **a**

   Answer: b) N/m Difficulty: 2

3. $\Sigma \mathbf{F} = 0$

   Answer: c) equilibrium Difficulty: 2

4. SI unit of force

   Answer: d) newton Difficulty: 2

5. coefficient of kinetic friction

   Answer: e) dimensionless Difficulty: 2

6. fps unit of mass

   Answer: f) slug Difficulty: 2

7. the amount of inertia

   Answer: g) mass Difficulty: 2

8. fps unit of force

   Answer: h) pound Difficulty: 2

9. cgs unit of acceleration

   Answer: i) cm/s^2 Difficulty: 2

Chapter 4

MULTIPLE CHOICE

10. Galileo's Law of Inertia is the same as
    a) Newton's First Law.
    b) Newton's Second Law.
    c) Newton's Third Law.
    d) Kepler's Law.

    Answer: a Difficulty: 2

11. You are standing in a moving bus, and you suddenly fall backward. You can imply from this that the bus's
    a) velocity increased.
    b) velocity decreased.
    c) speed remained the same, but its turning to the right.
    d) speed remained the same, but its turning to the left.

    Answer: a Difficulty: 2

12. Who has a greater weight to mass ratio, a person weighing 100 lb or a person weighing 150 lb?
    a) the person weighing 100 lb
    b) the person weighing 150 lb
    c) neither; their ratios are the same
    d) the question can't be answered; not enough information is given.

    Answer: c Difficulty: 2

13. Its more difficult to start moving a heavy carton from rest than it is to keep pushing it with constant velocity, because
    a) the normal force (N) is greater when the carton is at rest.
    b) $\mu_s < \mu_k$
    c) initially, the normal force (N) is not perpendicular to the applied force.
    d) $\mu_k < \mu_s$

    Answer: d Difficulty: 2

14. A packing crate slides down an inclined ramp at constant velocity. Thus we can deduce that
    a) a frictional force is acting on it.
    b) a net downward force is acting on it.
    c) it may be accelerating.
    d) it is not acted on by appreciable gravitational force.

    Answer: a Difficulty: 2

15. The number of forces acting on a car parked on a hill is
    a) one.
    b) two.
    c) three.
    d) four.

    Answer: b Difficulty: 2

16. You fall, while skiing, and one ski (of weight W) loosens and slides down an icy slope (assume no friction), which makes an angle θ with the horizontal. The force that pushes it down along the hill is
    a) zero; it moves with constant velocity.
    b) W
    c) W cos θ
    d) W sin θ

    Answer: d Difficulty: 2

17. Stacy (S) and David (D) are having a tug-of-war by pulling on opposite ends of a 5-kg rope. Stacy pulls with a 15-N force. What is David's force if the rope accelerates toward Stacy at 2 m/s^2?
    a) 3 N
    b) 5 N
    c) 25 N
    d) 50 N

    Answer: b Difficulty: 2

18. When you sit on a chair, the resultant force on you
    a) is zero.
    b) is up.
    c) is down.
    d) depends on your weight.

    Answer: a Difficulty: 2

TRUE/FALSE

19. A book can slide down a frictionless hill at constant velocity.

    Answer: F Difficulty: 2

Chapter 4

MULTIPLE CHOICE

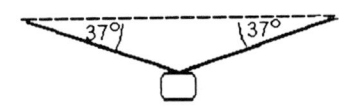

20. 1300-N gondola car, at a ski lift, is temporarily suspended at the halfway point, causing the wire to sag by 37° below the horizontal. The tension in the cable is
    a) 814 N
    b) 1080 N
    c) 1628 N
    d) 2160 N

    Answer: b Difficulty: 2

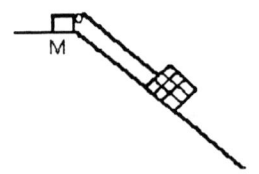

21. A rope, tied between a motor and a crate, can pull the crate up a hill of ice (assume no friction) at constant velocity. The free-body diagram of the crate should contain
    a) one force.
    b) two forces.
    c) three forces.
    d) four forces.

    Answer: c Difficulty: 2

22. The coefficients of friction for plastic on wood are $\mu_s = 0.5$ and $\mu_k = 0.4$. How much horizontal force would you need to apply to a 3.0-lb plastic calculator to start it moving from rest?
    a) 0.15 N
    b) 1.2 N
    c) 1.5 N
    d) 2.7 N

    Answer: c Difficulty: 2

Testbank

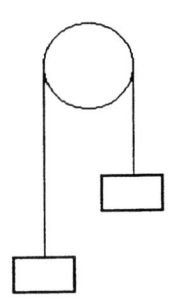

23. If two identical masses are attached by a light cord passing over a massless, frictionless pulley of an Atwood's machine, but at different heights, and then released,
    a) the lower mass will go down.
    b) the higher mass will go down.
    c) the masses will not move.
    d) the motion will depend on the amount of the masses.

    Answer: c Difficulty: 2

24. An astronaut is inside a space capsule in orbit around the earth. She is able to float inside the capsule because
    a) her weight is zero, and her capsule's weight is zero.
    b) her weight is zero, and her capsule is accelerating.
    c) she and her capsule move with the same constant velocity.
    d) she and her capsule move with the same constant acceleration.

    Answer: d Difficulty: 2

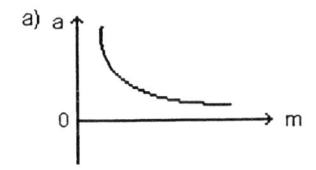

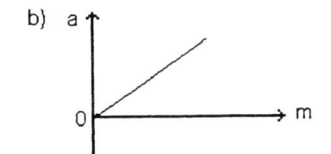

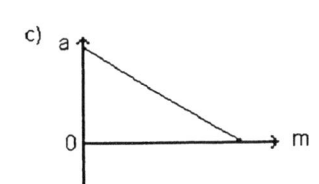

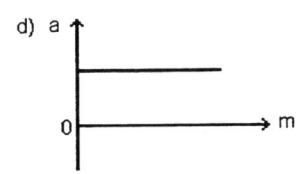

25. The same horizontal force is applied to objects of different mass. Which of the following graphs illustrates the experimental results?

    Answer: a Difficulty: 2

55

Chapter 4

26. If you blow up a balloon, and then release it, the balloon will fly away. This is an illustration of
   a) Newton's First Law.
   b) Newton's Second Law.
   c) Newton's Third Law.
   d) Galileo's Law of Inertia.

   Answer: c Difficulty: 2

27. When the rocket engines on the starship NO-PAIN-NO-GAIN are suddenly turned off, while traveling in empty space, the starship will
   a) stop immediately.
   b) slowly slow down, and then stop.
   c) go faster and faster.
   d) move with constant speed.

   Answer: d Difficulty: 2

28. If you exert a force F on an object, the force which the object exerts on you will
   a) depend on whether or not the object is moving.
   b) depend on whether or not you are moving.
   c) depend on the relative masses of you and the object.
   d) be F in all cases.

   Answer: d Difficulty: 2

29. Batter up! Your bat hits the ball pitched to you with a 1500-N instantaneous force. The ball hits the bat with an instantaneous force, whose magnitude is
   a) somewhat less than 1500 N.
   b) somewhat greater than 1500 N.
   c) exactly equal to 1500 N.
   d) essentially zero.

   Answer: c Difficulty: 2

ESSAY

30. A sports car of mass 1000 kg can accelerate from rest to 60 mi/h (=88ft/s) in 7 s. What would the average force of the car's engine be?

   Answer: 3830 N Difficulty: 2

56

MULTIPLE CHOICE

31. A child's toy is suspended from the ceiling by means of a string. The earth pulls downward on the toy with its weight force of 8 N. If this is the "action force," what is the "reaction force"?
    a) The string pulling upward on the toy with an 8-N force.
    b) The ceiling pulling upward on the string with an 8-N force.
    c) The string pulling downward on the ceiling with an 8-N force.
    d) The toy pulling upward on the earth with an 8-N force.

    Answer: d Difficulty: 2

TRUE/FALSE

32. Your mass on the moon will be about one-sixth of your mass on earth.

    Answer: F Difficulty: 2

MULTIPLE CHOICE

33. An arrow is shot straight up. At the top of its path, the net force acting on it is
    a) greater than its weight.
    b) greater than zero, but less than its weight.
    c) instantaneously equal to zero.
    d) equal to its weight.

    Answer: d Difficulty: 2

ESSAY

34. A sled of mass 10 kg slides down a flat hill that makes an angle of 10° with the horizontal. If friction is negligible, what is the resultant force on the sled?

    Answer: 17.0 N Difficulty: 2

MULTIPLE CHOICE

35. Two toy cars (16 kg and 2 kg) are released simultaneously on an inclined plane that makes an angle of 30° with the horizontal. Which statement best describes their acceleration after being released?
    a) The 16-kg car accelerates 8 times faster than the 2-kg car.
    b) The 2-kg car accelerates 8 times faster than the 16-kg car.
    c) Both cars accelerate at a rate of 0.866 g.
    d) None of the above.

    Answer: b Difficulty: 2

Chapter 4

36. A block of mass M slides down a frictionless plane inclined at an angle θ with the horizontal. The normal reaction force exerted by the plane on the block is
a) Mg
b) Mg sin θ
c) Mg cos θ
d) zero, since the plane is frictionless.

Answer: c Difficulty: 2

37. Florence, who weighs 120 lb, stands on a bathroom scale in an elevator. What will she see the scale read when the elevator is accelerating upward at 4.0 m/s^2?
a) 120 lb
b) 135 lb
c) 105 lb
d) 90 lb

Answer: b Difficulty: 2

38. A decoration, of mass M, is suspended by a string from the ceiling inside an elevator. The elevator is traveling upward with a constant speed. The tension in the string is
a) equal to Mg.
b) less than Mg.
c) greater than Mg.
d) impossible to tell without knowing the speed.

Answer: a Difficulty: 2

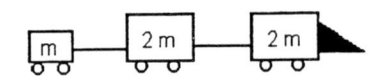

[Figure: three cars labeled m, 2m, 2m on wheels with incline on right]

39. A train consists of a caboose (mass = 1000 kg), a car (mass 2000 kg), and an engine car (mass 2000 kg). If the train has an acceleration of 5 m/s^2, then the tension force in the coupling between the middle car and the engine car is
a) 25,000 N
b) 20,000 N
c) 15,000 N
d) 10,000 N

Answer: c Difficulty: 2

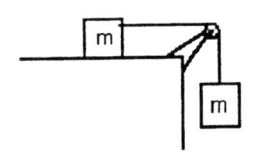

40. Two identical masses are attached by a light string that passes over a small pulley, as shown. The table and the pulley are frictionless. The system is moving
    a) with an acceleration less than g.
    b) with an acceleration equal to g.
    c) with an acceleration greater than g.
    d) at constant speed.

    Answer: a Difficulty: 2

41. If you push a 4-kg mass with the same force that you push a 10-kg mass from rest,
    a) the 10-kg mass accelerates 2.5 times faster than the 4-kg mass.
    b) the 4-kg mass accelerates 2.5 times faster than the 10-kg mass.
    c) both masses accelerate at the same rate.
    d) none of the above is true.

    Answer: b Difficulty: 2

42. A 4-kg mass and a 10-kg mass are acted on by the same constant net force during the same amount of time. Both masses are at rest before the force is applied. The 10-kg mass moves a distance X_1 and the 4-kg mass moves a distance X_2 as a result. Which one of the following statements is true?
    a) X_1 is equal to X_2.
    b) The ratio X_1/X_2 is equal to 5/2.
    c) The ratio X_1/X_2 is equal to 2/5.
    d) The ratio X_1/X_2 is equal to $(2/5)^2$.

    Answer: b Difficulty: 2

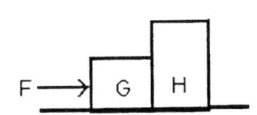

43. Two cardboard boxes full of books are in contact with each other on a table. Box H has twice the mass of box G. If you push on box G with a horizontal force F, then box H will experience a net force of
    a) 2/3 F
    b) F
    c) 3/2 F
    d) 2 F

    Answer: a Difficulty: 2

Chapter 4

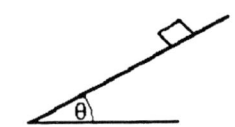

44. A toolbox, of mass M, is resting on a flat board. One end of the board is lifted up until the toolbox just starts to slide. The angle θ that the board makes with the horizontal, for this to occur, depends on the
    a) mass, M.
    b) acceleration of gravity, g.
    c) normal force.
    d) coefficient of static friction, μ_s.

    Answer: d Difficulty: 2

TRUE/FALSE

45. The angle of repose depends on the weight of the object on the hill.

    Answer: F Difficulty: 2

MULTIPLE CHOICE

46. A horizontal force of 5 N accelerates a 4-kg mass, from rest, at a rate of 0.5 in the positive direction. What friction force acts on the mass?
    a) +3 N
    b) -3 N
    c) +2 N
    d) -2 N

    Answer: a Difficulty: 2

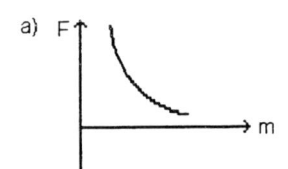

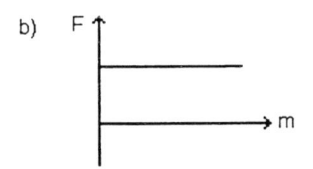

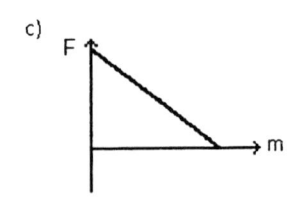

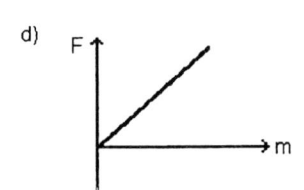

47. Four students perform an experiment by pulling an object horizontally across a frictionless table. They repeat the experiment on several other objects of different masses, always applying just enough force to produce the same acceleration. They each graph the results, individually. Which student's graph is correct?

    Answer: d Difficulty: 2

48. A horizontal force accelerates a box from rest across a horizontal surface (friction is present) at a constant rate. All conditions remain the same with the exception that the horizontal force is doubled. What happens to the box's acceleration?
a) It increases to more than double its original value.
b) It increases to exactly double its original value.
c) It increases to less than double its original value.
d) It increases somewhat.

Answer: a Difficulty: 2

ESSAY

49. A 2-kg mass and a 5-kg mass are hung on opposite sides of an Atwood's machine, and released. (a) What is the tension in the string? (b) How much time is required to fall 60 cms from rest?

Answer: (a) 16.8 N (b) 0.53 s Difficulty: 2

50. A load of steel of mass 6000 kg rests on the flatbed of a truck. It is held in place by metal brackets that can exert a maximum horizontal force of 8000 N. When the truck is traveling 20 m/s, what is the minimum stopping distance if the load is not to slide forward into the cab?

Answer: 150 m Difficulty: 2

MULTIPLE CHOICE

51. An antitank weapon fires a 3-kg rocket which acquires a speed of 50 m/s after traveling 90 cm down a launching tube. Assuming the rocket was accelerated uniformly, what force acted on it?
a) 4170 N
b) 3620 N
c) 2820 N
d) 2000 N

Answer: a Difficulty: 2

ESSAY

52. During the investigation of a traffic accident, police find skid marks 90 m long. They determine the coefficient of friction between the car's tires and the roadway to be 0.5 for the prevailing conditions. Estimate the speed of the car when the brakes were applied.

Answer: 29.7 m/s Difficulty: 2

Chapter 4

53. A bulldozer drags a log weighing 500 N along a rough surface. The cable attached to the log makes an angle of 30° with the ground. The coefficient of static friction between the log and the ground is 0.5. What minimum tension is required in the cable in order for the log to begin to move?

    Answer: 224 N Difficulty: 2

MULTIPLE CHOICE

54. A 16-kg fish is weighed with two spring scales, each of negligible weight, as shown here. What will be the readings on the scales?
    a) Each scale will read 8 kg.
    b) Each scale will read 16 kg.
    c) The top scale will read 16 kg, and the bottom scale will read zero.
    d) The bottom scale will read 16 kg, and the top scale will read zero.
    e) Each scale will show a reading greater than zero and less then 16 kg, but the sum of the two readings will be 16 kg.

    Answer: b Difficulty: 2

55. A brick and a feather fall to the earth at their respective terminal velocities. Which object experiences the greater force of air friction?
    a) The feather
    b) The brick
    c) Neither, both experience the same amount of air friction.

    Answer: b Difficulty: 2

56. A wooden block slides directly down an inclined plane, at a constant velocity of 6 m/s. How much is the coefficient of kinetic friction (μ_k), if the plane makes an angle of 25° with the horizontal?
    a) 0.47
    b) 0.42
    c) 0.37
    d) 0.91
    e) 2.1

    Answer: a Difficulty: 2

ESSAY

57. A flatbed truck carries a load of steel. Only friction keeps the steel from sliding on the bed of the truck. The driver finds that the minimum distance in which he can stop from a speed of 20 mi/h without having the steel slide forward into the truck cab is 20 m. What would his minimum stopping distance be when going down a 10° incline?

Answer: 153 m   Difficulty: 2

MULTIPLE CHOICE

58. Suppose several forces are acting upon a mass m. The F in F=ma refers to
    a) the arithmetic sum of all the forces
    b) the vector sum of all the forces
    c) to any particular one of the forces
    d) the force of friction acting upon the mass

Answer: b   Difficulty: 1

59. An object is dropped from a helicoptor. When it reaches terminal velocity
    a) gravity no longer acts upon it
    b) all of the upward forces add up to zero
    c) the acceleration finally reaches g
    d) the total force is zero

Answer: d   Difficulty: 1

60. The English pound is equivalent to about how many Newtons?
    a) 1
    b) 2.2
    c) 4.5
    d) 0.22
    e) 0.04

Answer: c   Difficulty: 1

61. The kinetic friction of one object sliding along another surface depends least upon:
    a) the normal force
    b) the area of contact
    c) the coefficient of kinetic friction
    d) the nature of the materials in contact

Answer: b   Difficulty: 1

Chapter 4

62. The statement by Newton that "for every action there is an opposite but equal reaction" is regarded as which of his laws of motion?
   a) first
   b) second
   c) third
   d) fourth
   e) gravitation

   Answer: c Difficulty: 1

63. Work done by **STATIC FRICTION** is always:
   a) negative
   b) positive
   c) along the surface
   d) zero

   Answer: d Difficulty: 1

64. An elevator weighing 10,000.N is supported by a steel cable. Determine the tension in the cable when the elevator is accelerated upward at 3.0 m/s^2 (g=9.8 m/s^2)
   a) 7.0 kN
   b) 10.0 kN
   c) 11.6 kN
   d) 13.1 kN
   e) 40.0 kN

   Answer: d Difficulty: 2

65. Planning to elope with her boyfriend, Janice slides down a rope from her fourth story window. As she slides down the rope, she tightens harder on the rope which of course increases the tension in the rope. When the upward tension in the rope equals Janice's weight:
   a) Janice will slow down
   b) Janice will speed up
   c) Janice will continue down a constant speed
   d) Janice will speed up
   e) the rope will break.

   Answer: c Difficulty: 2

66. Which is usually larger?
   a) kinetic coefficient of friction
   b) static coefficient of friction

   Answer: b Difficulty: 1

FILL-IN-THE-BLANK

67. The MKS unit of force is a _____ and the CGS force unit is a _____ and the English unit for force is a _____.

    Answer: Newton, Dyne, Pound   Difficulty: 1

ESSAY

68. How important is Newton's third law? If nature did not obey Newton's #3 would our application of Newton's 2nd law be much different?

    Answer:
    When we apply F=ma to any complex system (group of atoms or masses), we can ignore internal forces because they cancel each other in pairs. If there were not an "action-reaction" law, it would be enormously more difficult (if not impossible) to analyse macroscopic systems (imagine needing to know the internal forces on each of the 6×10^{26} molecules in a kg-mole).
    Difficulty: 2

FILL-IN-THE-BLANK

69. A 170. pound halfback is tackled by a 245. pound guard. Each, of course, exerts a force on the other. Which of the players exerts the LARGER force on the other?

    Answer: neither   Difficulty: 2

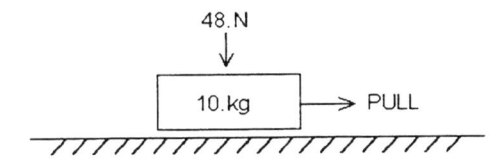

70. A 10.kg blocks rests on a table (on Earth) and 0.5 is the coefficient of static friction. If someone pushes down on the block with a 48.Newton force, what is the maximum horizontal pull (P) that will not move the block?

    Answer: 73.Newtons   Difficulty: 2

Chapter 4

MULTIPLE CHOICE

71. An average sized apple weighs how many Newtons?
   a) 2.
   b) 200.
   c) 40.
   d) 0.04
   e) 0.004

   Answer: a   Difficulty: 1

FILL-IN-THE-BLANK

72. What is the weight of a 14.0 kg pumpkin?

   Answer: mg=137.Newtons   Difficulty: 1

MULTIPLE CHOICE

73. If a forces accelerates 4.5 kg at 40.m/s^2, that same force would accelerate 18.kg by how much?
   a) 0.16 km/s^2
   b) 10. m/s^2
   c) 0.18 km/s^2
   d) 40. m/s^2
   e) 32. ft/s^2

   Answer: b   Difficulty: 2

ESSAY

74. State Newton's 3rd law in your own words.

   Answer:
   If one thing pushes another, the other pushes back equally (opposite direction); the "action-reaction law"

   Difficulty: 2

FILL-IN-THE-BLANK

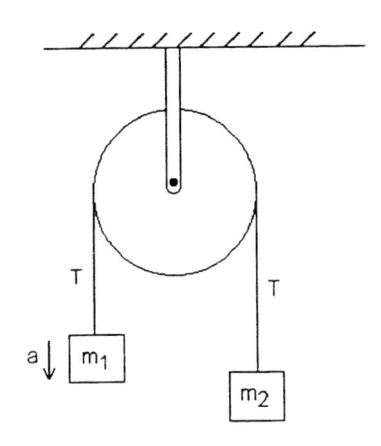

75. Consider the Atwood machine in the accompanying figure. m_1=33.0kg and m_2=11.0kg and the pulley mass is negligible.
    a) Determine the acceleration.
    b) What is the tension in the connecting string?

    Answer: a) 4.90 m/s^2 b) tension= 162.Newtons   Difficulty: 2

MULTIPLE CHOICE

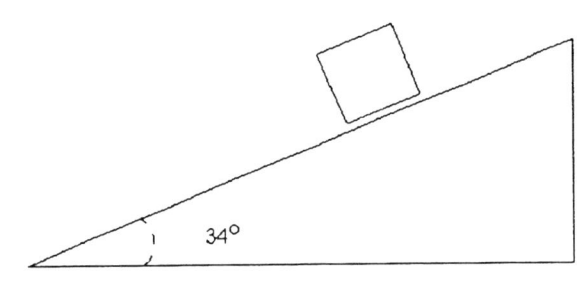

76. A 42.0kg block of ice slides down the plane incline 34.0° in the figure. Assuming friction is negligible, what is the acceleration of the block down the incline?

    a) 5.48 m/s^2
    b) 8.12 m/s^2
    c) 9.80 m/s^2
    d) 6.61 m/s^2

    Answer: a   Difficulty: 2

Chapter 4

77. Consider the 42.0kg block of ice illustrated in problem 4-19. Assuming friction that the kinetic coefficient of friction is 0.060, what is the acceleration of the block down the incline?

   a) 5.43 m/s^2
   b) 8.06 m/s^2
   c) 9.74 m/s^2
   d) 6.56 m/s^2

   Answer: a   Difficulty: 2

Chapter 5

MATCHING

Match the unit or expression to the physical quantity.

a) 3.6 MJ
b) N·m
c) dyne·cm
d) ft·lb
e) 1/2 mv²

f) mgy
g) weight
h) friction
i) N/m
j) 550 ft·lb/s

k) dimensionless
l) -kx
m) J/s
n) ft·lb/s

1. kWh

   Answer: a) 3.6 MJ Difficulty: 2

2. joule

   Answer: b) N·m Difficulty: 2

3. erg

   Answer: c) dyne·cm Difficulty: 2

4. fps unit of work

   Answer: d) ft·lb Difficulty: 2

5. kinetic energy

   Answer: e) 1/2 mv² Difficulty: 2

6. potential energy

   Answer: f) mgy Difficulty: 2

7. conservative force

   Answer: g) weight Difficulty: 2

8. non-conservative force

   Answer: h) friction Difficulty: 2

9. spring constant

   Answer: i) N/m Difficulty: 2

Chapter 5

10. hp

    Answer: j) 550 ft·lb/s   Difficulty: 2

11. efficiency

    Answer: k) dimensionless   Difficulty: 2

12. Hooke's Law

    Answer: l) $-kx$ Difficulty: 2

13. watt

    Answer: m) J/s   Difficulty: 2

14. fps unit of power

    Answer: n) ft·lb/s   Difficulty: 2

MULTIPLE CHOICE

15. Two men, Joel and Jerry, push against a wall. Jerry stops after 10 min, while Joel is able to push for 5 min longer. Compare the work they do.
    a) Joel does 50% more work than Jerry.
    b) Jerry does 50% more work than Joel.
    c) Joel does 75% more work than Jerry.
    d) Neither of them do any work.

    Answer: d   Difficulty: 2

16. You lift a 10-lb physics book up in the air a distance of 1 ft, at a constant velocity of 0.5 ft/s. The work done by gravity is
    a) +10 ft·lb
    b) -10 ft·lb
    c) +5 ft·lb
    d) -5 ft·lb
    e) zero

    Answer: b   Difficulty: 2

17. The area under the curve, on a Force-position (F-x) graph, represents
    a) work.
    b) kinetic energy.
    c) power.
    d) efficiency.

    Answer: a   Difficulty: 2

a) [graph: F constant positive vs x]
b) [graph: F linear negative slope vs x]
c) [graph: F increasing curve vs x]
d) [graph: F decreasing curve vs x]

18. Which of the following graphs illustrates Hooke's Law?

    Answer: b Difficulty: 2

19. A 10-kg mass, hung onto a spring, causes the spring to stretch 2 cm. The spring constant is
    a) 49 N/cm
    b) 5 N/cm
    c) 0.2 N/cm
    d) 0.02 N/cm

    Answer: a Difficulty: 2

20. Car J moves twice as fast as car K, and car J has half the mass of car K. The kinetic energy of car J, compared to car K is
    a) the same.
    b) 2 to 1.
    c) 4 to 1.
    d) 1 to 2.

    Answer: b Difficulty: 2

21. A leaf falls from a tree. Compare its kinetic energy K, to its potential energy U.
    a) K increases, and U decreases.
    b) K decreases, and U decreases.
    c) K increases, and U increases.
    d) K decreases, and U increases.

    Answer: a Difficulty: 2

Chapter 5

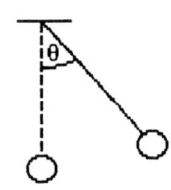

22. A simple pendulum, consisting of a mass m and a string, swings upward, making an angle θ with the vertical. The work done by the tension force is
    a) zero
    b) mg
    c) mg cos θ
    d) mg sin θ

    Answer: a Difficulty: 2

23. Matthew pulls his little sister Sarah in a sled on an icy surface (assume no friction), with a force of 60 N at an angle of 37° upward from the horizontal. If he pulls her a distance of 12 m, the work he does is
    a) 185 J
    b) 433 J
    c) 575 J
    d) 720 J

    Answer: c Difficulty: 2

ESSAY

24. A driver, traveling at 66 ft/s, slows down her 3000-lb car to stop for a red light. What work is done by the friction force against the wheels?

    Answer: 2.04 x 10^5 ft·lb Difficulty: 2

MULTIPLE CHOICE

25. A 4.0-kg box of fruit slides 8.0 m down a ramp, inclined at 30° from the horizontal. If the box slides at a constant velocity of 5 m/s, the work done by gravity is
    a) zero
    b) +78 J
    c) -78 J
    d) +157 J
    e) -157 J

    Answer: d Difficulty: 2

26. A 200-g mass attached to the end of a spring causes it to stretch 5.0 cm. If another 200-g mass is added to the spring, the potential energy of the spring will be
a) the same.
b) twice as much.
c) 3 times as much.
d) 4 times as much.

Answer: d Difficulty: 2

27. You throw a ball straight up. Compare the sign of the work done by gravity while the ball goes up with the sign of the work done by gravity while it goes down.
a) Work up is +, and the work down is +.
b) Work up is +, and the work down is -.
c) Work up is -, and the work down is +.
d) Work up is -, and the work down is -.

Answer: c Difficulty: 2

NOTE: The following question(s) refers to the Cyclone, the famous roller coaster ride at Coney sland, shown in the sketch. Assume no friction.)

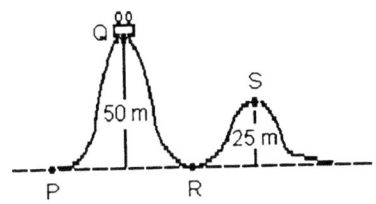

28. How much work was required to bring the 1000-kg roller coaster from point P to rest at point Q at the top of the 50 m peak?
a) 32,000 J
b) 50,000 J
c) 245,000 J
d) 490,000 J

Answer: d Difficulty: 2

29. If the roller coaster leaves point Q from rest, how fast is it traveling at point R?
a) 22.1 m/s
b) 31.3 m/s
c) 490 m/s
d) 980 m/s

Answer: b Difficulty: 2

Chapter 5

30. If the roller coaster leaves point Q from rest, what is its speed at point S (at the top of the 25-m peak) compared to its speed at point R?
    a) zero
    b) $1/\sqrt{2}$
    c) $\sqrt{2}$
    d) 2
    e) 4

Answer: b Difficulty: 2

31. A container of water is lifted vertically 3 m, then returned to its original position. If the total weight is 30 N, how much work was done?
    a) 45 J
    b) 90 J
    c) 180 J
    d) No work was done.

Answer: d Difficulty: 2

32. On a plot of F vs. x, what represents the work done by the force F?
    a) The slope of the curve
    b) The length of the curve
    c) The area under the curve
    d) The product of the maximum force times the maximum x

Answer: c Difficulty: 2

33. A 1-kg flashlight falls to the floor. At the point during its fall, when it is 0.70 m above the floor, its potential energy exactly equals its kinetic energy. How fast is it moving?
    a) 3.7 m/s
    b) 6.9 m/s
    c) 13.7 m/s
    d) 44.8 m/s

Answer: a Difficulty: 2

34. A truck weighs twice as much as a car, and is moving at twice the speed of the car. Which statement is true about the truck's kinetic energy (K) compared to that of the car?
    a) All that can be said is that the truck has more K.
    b) The truck has twice the K of the car.
    c) The truck has 4 times the K of the car.
    d) The truck has 8 times the K of the car.

Answer: d Difficulty: 2

35. Is it possible for a system to have negative potential energy?
   a) Yes, as long as the total energy is positive.
   b) Yes, since the choice of the zero of potential energy is arbitrary.
   c) No, because the kinetic energy of a system must equal its potential energy.
   d) No, because this would have no physical meaning.

   Answer: b Difficulty: 2

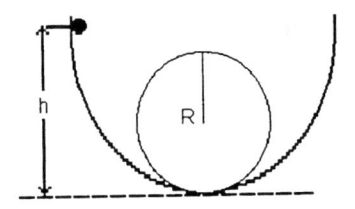

36. (LOOP-THE-LOOP) A ball is released, from rest, at the left side of the loop-the-loop, at the height shown. If the radius of the loop is R, what vertical height does the ball rise to, on the other side, neglecting friction?
   a) Less than R
   b) R
   c) 2R
   d) Greater than 2R

   Answer: d Difficulty: 2

37. A spring is characterized by a spring constant of 60 N/m. How much potential energy does it store, when stretched by 1 cm?
   a) 0.003 J
   b) 0.3 J
   c) 60 J
   d) 600 J

   Answer: a Difficulty: 2

38. You and your friend want to go to the top of the Eiffel Tower. Your friend takes the elevator straight up. You decide to walk up the spiral stairway, taking longer to do so. Compare the gravitational potential energy (U) of you and your friend, after you both reach the top.
   a) Your friend's U is greater than your U, because she got to the top faster.
   b) Your U is greater than your friend's U, because you traveled a greater distance in getting to the top.
   c) Both of you have the same amount of potential energy.
   d) It is impossible to tell, since the times and distances are unknown.

   Answer: c Difficulty: 2

Chapter 5

39. The total mechanical energy of a system
    a) is equally divided between kinetic energy and potential energy.
    b) is either all kinetic energy or all potential energy, at any one instant.
    c) can never be negative.
    d) is constant, only if conservative forces act.

    Answer: d Difficulty: 2

40. A skier, of mass 40 kg, pushes off the top of a hill with an initial speed of 4 m/s. How fast will she be moving after dropping 10 m in elevation?
    a) 7 m/s
    b) 14 m/s
    c) 49 m/s
    d) 196 m/s

    Answer: b Difficulty: 2

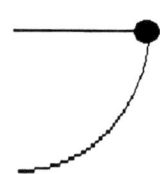

41. A simple pendulum, consisting of a mass m, is attached to the end of a 1 yd length of string. If the mass is held out horizontally, and then released from rest, its speed at the bottom is
    a) 9.8 m/s
    b) 14 m/s
    c) 96 m/s
    d) 288 m/s

    Answer: b Difficulty: 2

42. A toy rocket, weighing 10 N, blasts off from ground level with a kinetic energy of 40 J. At the exact top of its trajectory, its energy is 140 J. To what vertical height does it rise?
    a) 10 m
    b) 12 m
    c) 22 m
    d) None of the above

    Answer: a Difficulty: 2

43. A boy releases his 2-kg toy, from rest, at the top of a sliding-pond inclined at 20° above the horizontal. What will the toy's speed be after sliding 4 m along the sliding-pond? The coefficient of kinetic friction is 0.2.
    a) 2.21 m/s
    b) 3.00 m/s
    c) 3.48 m/s
    d) 5.18 m/s

    Answer: d Difficulty: 2

44. A 4-kg mass moving with speed 2 m/s, and a 2-kg mass moving with a speed of 4 m/s, are gliding over a horizontal frictionless surface. Both objects encounter the same horizontal force, which directly opposes their motion, and are brought to rest by it. Which statement best describes their respective stopping distances?
    a) The 4-kg mass travels twice as far as the 2-kg mass before stopping.
    b) The 2-kg mass travels twice as far as the 4-kg mass before stopping.
    c) Both masses travel the same distance before stopping.
    d) The 2-kg mass travels farther, but not necessarily twice as far.

    Answer: b Difficulty: 2

45. King Kong falls from the top of the Empire State Building, through the air (air friction is present), to the ground below. How does his kinetic energy (K) just before striking the ground compare to his potential energy (U) at the top of the building?
    a) K is equal to U.
    b) K is greater than U.
    c) K is less than U.
    d) It is impossible to tell.

    Answer: c Difficulty: 2

46. You slam on the brakes of your car in a panic, and skid a certain distance on a straight, level road. If you had been traveling twice as fast, what distance would the car have skidded, under the same conditions?
    a) It would have skidded 4 times farther.
    b) It would have skidded twice as far.
    c) It would have skidded $\sqrt{2}$ times farther.
    d) It is impossible to tell from the information given.

    Answer: a Difficulty: 2

ESSAY

47. How many joules of energy are used by a 1 hp motor that runs for 1 hr?

    Answer: 2.69×10^6 J Difficulty: 2

Chapter 5

48. Water flows over a waterfall 20 m high, at the rate of 2 x 10⁴ kg/s. If this water is used to power an electric generator with a 40% efficiency, how many joules of electric energy can be supplied?

    Answer: 1.57 x 10⁶ J Difficulty: 2

49. A roofer lifts supplies a height of 20 m with a hand-operated winch. How long would it take him to lift a 200-kg load, if the winch has an efficiency of 50%, and the rate at which the winch can do work is 0.8 hp?

    Answer: 131 s Difficulty: 2

50. A cyclist does work at the rate of 500 W while riding. How much force does her foot push with when she is traveling at 8 m/s?

    Answer: 62.5 N Difficulty: 2

51. Assuming muscles are 20% efficient, at what rate is a 60-kg boy using energy when he runs up a flight of stairs 10 m high, in 8 s?

    Answer: 735 W Difficulty: 2

MULTIPLE CHOICE

52. To accelerate your car at a constant acceleration, the car's engine must
    a) maintain a constant power output.
    b) develop ever-decreasing power.
    c) develop ever-increasing power.
    d) maintain a constant turning speed.

    Answer: c Difficulty: 2

53. Compared to yesterday, you did 3 times the work in one-third the time. To do so, your power output must have been
    a) the same as yesterday's power output.
    b) one-third of yesterday's power output.
    c) 3 times yesterday's power output.
    d) 9 times yesterday's power output.

    Answer: d Difficulty: 2

54. A brick and a pebble fall from the roof of an apartment building under construction. At some point the brick is moving at a speed of 3 m/s and the pebble's speed is 5 m/s. If both objects have the same kinetic energy, what is the ratio of the brick's mass to the rock's mass?
    a) 25 to 9
    b) 5 to 3
    c) 12.5 to 4.5
    d) 3 to 5

    Answer: a Difficulty: 2

55. A 30-N stone is dropped from a height of 10 m, and strikes the ground with a velocity of 7 m/s. What average force of air friction acts on it as it falls?
a) 22.5 N
b) 75 N
c) 225 N
d) 293 N

Answer: a Difficulty: 2

ESSAY

56. A family goes on vacation for one week, but forgets to turn off an electric fan that consumes electricity at the rate of 200 Watts. If the cost of electricity is 12 /kWh, how much money does it cost to run the fan?

Answer: $4.03 Difficulty: 2

MULTIPLE CHOICE

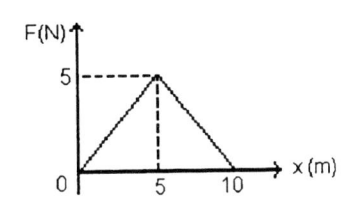

57. The force that a spider exerts on the fly it's captured is observed over a 10-second interval, as shown on the graph. How much work did the spider do during that 10 s?
a) Zero
b) 12.5 J
c) 25 J
d) 50 J

Answer: c Difficulty: 2

Chapter 5

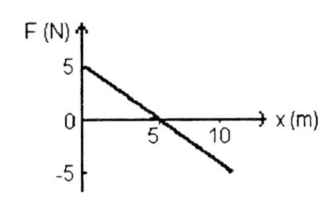

58. The resultant force you exert while pressing a key on the keyboard of your new computer, for a 1.0-s period, is plotted on the graph, shown. How much work did you do during this 1-s interval?
a) Zero
b) 12.5 J
c) -25 J
d) 50 J

Answer: a   Difficulty: 2

59. When you lift a 12. ounce beverage can from the table top to your mouth, you do approximately how much work?
a) 1 Calorie
b) 1 Watt
c) 1 Kw-h
d) 1 Joule
e) 1 erg

Answer: d   Difficulty: 1

60. A "machine" multiples (increases):
a) work
b) energy
c) time
d) force

Answer: d   Difficulty: 1

FILL-IN-THE-BLANK

61. Daisy raises 10.Kg to a height of 2.5 meters in 2.0 seconds.

a) How much work did she do? _____

b) How much power was expended? _____

c) If she raises it in 1.0 s rather than 2.0s, how does the work and power change?

Answer: a) 0.25 KJ   b) 0.12 Kw   c) same work, but power doubles

Difficulty: 2

TRUE/FALSE

62. Kinetic energy is proportional to speed.

    Answer: F Difficulty: 1

MULTIPLE CHOICE

63. A horse-power is equal to 550:
    a) J/hour
    b) ft-lb/minute
    c) Kw-h
    d) ft-lb/s
    e) J/s

    Answer: d Difficulty: 1

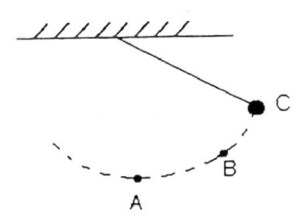

64. On the accompanying diagram of a pendulum, at what position is the potential energy maximum?
    a) A
    b) B
    c) C

    Answer: c Difficulty: 1

65. On the accompanying diagram of a pendulum, at what position is the kinetic energy maximum?
    a) A
    b) B
    c) C

    Answer: a Difficulty: 1

66. Which of the following is **not** a unit of work?
    a) Kg-m/s
    b) J
    c) Kw-h
    d) N-m
    e) W-s

    Answer: a Difficulty: 1

Chapter 5

67. If both mass and the velocity of a ball are each tripled, the kinetic energy is increased by a factor of:
a) 3
b) 6
c) 9
d) 18
e) 27
f) 81

Answer: e Difficulty: 2

FILL-IN-THE-BLANK

68. Place a P or an E after each unit, indicating whether it is a unit of Power or Energy:

watt____   ft-lb/s____   joule____

erg ____   Newton-meter____   ft-lb____

kw-h____   horsepower____   calorie____

watt-s____   kg-m^2/s^2____

Answer:
P P E
E E E
E P E
E P

Difficulty: 1

MULTIPLE CHOICE

69. If there is no motion, can work be done on a system?
a) Yes, provided an outside force is applied.
b) Yes, since motion is only relative.
c) No, since a system which is not moving has no kinetic energy.
d) No, because of the way work is defined.

Answer: d Difficulty: 2

70. My battery charger uses 12.watts. At 6.0 cents per kilowatt-hour, how much does it cost to charge batteries for 24.hours?
a) 0.28 cents
b) 1.4 cents
c) 0.74 cents
d) 1.7 cents
e) 2.3 cents

Answer: d Difficulty: 2

FILL-IN-THE-BLANK

71. Is more work required to increase a car's speed from 10.mph to 30.mph, or from 50.mph to 60.mph?

    Answer: 50 to 60 mph Difficulty: 2

72. An auto is coasting on a level road. It weighs 10. kN. How much work is done by gravity as it moves horizontally 150.meters ?

    Answer: zero (displacement is perpendicular to the force) Difficulty: 1

MULTIPLE CHOICE

73. Consider an ideal elastic spring. The spring constant is:
    a) proportional to displacement from equilibrium
    b) proportional to the applied force
    c) inversely proportional to the applied force
    d) proportional to the mass attached to the spring
    e) inversely proportioal to the displacement
    f) none of the above

    Answer: f Difficulty: 1

74. Consider a plot of the displacement (x) vs applied force (F) for an ideal elastic spring. The slope of the curve would be:
    a) the spring constant
    b) the reciprocal of the spring constant
    c) the acceleration of gravity
    d) the reciprocal of the acceleration of gravity
    e) none of the above

    Answer: b Difficulty: 2

75. The work energy theorem says
    a) the net work done is equal to the initial kinetic energy less the final energy.
    b) final kinetic energy plus the net work done is the initial kinetic energy.
    c) the net work done plus the initial kinetic energy is the final kinetic energy.
    d) the net work done plus the final kinetic energy is the initial kinetic energy.
    e) the net work done minus the initial kinetic energy is the final kinetic energy.

    Answer: c Difficulty: 2

Chapter 5

ESSAY

76. What distinguishes a conservative force from a non-conservative force?

    Answer:
    Work done by a conservative force is independent of the object's path and the work only depends upon the initial and final positions.
    Difficulty: 2

TRUE/FALSE

77. Problems involving a non-conservative force cannot be solved because there is no defineable potential energy.

    Answer: F Difficulty: 1

FILL-IN-THE-BLANK

78. As plant engineer, Donna purchases an electrical motor which has an output of 1500.hp. If it requires 1.20 Megawatts of electricity, what is the efficiency of the motor?

    Answer: 0.933 = 93.3 % Difficulty: 2

Chapter 6

MATCHING

Match the physical unit to the physical quantity.

a) slug-ft/s   d) newton
b) N-s   e) centimeter
c) pound

1. fps unit of impulse

   Answer: a) slug-ft/s Difficulty: 2

2. SI unit of momentum

   Answer: b) N-s Difficulty: 2

3. fps unit of $\Delta p/\Delta t$

   Answer: c) pound Difficulty: 2

4. SI unit of $\Delta p/\Delta t$

   Answer: d) newton Difficulty: 2

5. metric unit of center of mass

   Answer: e) centimeter Difficulty: 2

MULTIPLE CHOICE

6. A 1200-kg ferryboat is moving south at 20 m/s. Its momentum is
   a) 1.7×10^{-3} N·s
   b) 600 kg·m/s
   c) 2.4×10^3 kg·m/s
   d) 2.4×10^4 N·s

   Answer: d Difficulty: 2

7. When a light beach ball rolling with a speed of 6 m/s collides with a heavy exercise ball at rest, the beach ball's speed after the collision will be, approximately,
   a) 0
   b) 3 m/s
   c) 6 m/s
   d) 12 m/s

   Answer: d Difficulty: 2

Chapter 6

8. In a game of pool, the white cue ball hits the #5 ball and stops, while the #5 ball moves away with the same velocity as the cue ball had originally. The type of collision is
   a) elastic.
   b) inelastic.
   c) completely inelastic.
   d) any of the above, depending on the mass of the balls.

   Answer: a Difficulty: 2

TRUE/FALSE

9. It is physically impossible to make a 90° turn.

   Answer: T Difficulty: 2

MULTIPLE CHOICE

10. Tightrope walkers walk with a long flexible rod in order to
    a) increase their total weight.
    b) allow both hands to hold onto something.
    c) lower their center of mass.
    d) move faster along the rope.

    Answer: c Difficulty: 2

TRUE/FALSE

11. The center of mass of an object must always be located where the physical material of the object is located.

    Answer: F Difficulty: 2

MULTIPLE CHOICE

12. A 2-kg softball is pitched to you at 20 m/s. You hit the ball back along the same path, and at the same speed. If the bat was in contact with the ball for 0.1 s, the average force the bat exerted was
    a) zero
    b) 40 N
    c) 400 N
    d) 800 N

    Answer: d Difficulty: 2

13. The area under the curve on a F - t graph represents
    a) impulse.
    b) momentum.
    c) work.
    d) kinetic energy.

    Answer: b Difficulty: 2

14. You (50-kg mass) skate on ice at 4 m/s to greet your friend (40-kg mass), who is standing still, with open arms. As you collide, while holding each other, with what speed do you both move off together?
    a) Zero
    b) 2.2 m/s
    c) 5 m/s
    d) 22.5 m/s

    Answer: b Difficulty: 2

TRUE/FALSE

15. Momentum is conserved during a completely inelastic collision.

    Answer: T Difficulty: 2

MULTIPLE CHOICE

16. A handball of mass 0.1 kg, traveling horizontally at 30 m/s, strikes a wall and rebounds at 24 m/s. What is the change in the momentum of the ball?
    a) 0.6 m/s
    b) 1.2 m/s
    c) 5.4 m/s
    d) 72 m/s

    Answer: a Difficulty: 2

17. When a cannon fires a cannonball, the cannon will recoil backward because the
    a) energy of the cannonball and cannon is conserved.
    b) momentum of the cannonball and cannon is conserved.
    c) energy of the cannon is greater than the energy of the cannonball.
    d) momentum of the cannon is greater than the energy of the cannonball.

    Answer: b Difficulty: 2

Chapter 6

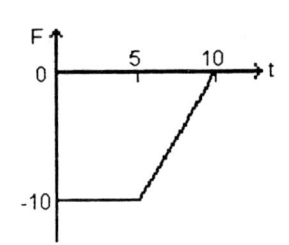

[Graph: F vs t, showing F=-10 from t=0 to t=5, then linearly rising to 0 at t=10]

18. A crane lifts up a 100-kg box, from rest, with a force that varies with time, as shown on the graph. At t = 10 s, the instantaneous velocity of the box is
    a) Zero
    b) +0.25 m/s
    c) -0.25 m/s
    d) +0.75 m/s
    e) -0.75 m/s

    Answer: e Difficulty: 2

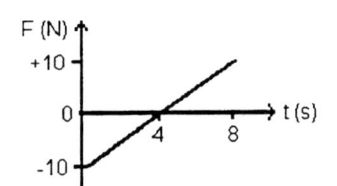

[Graph: F(N) vs t(s), linear line from -10 at t=0, crossing zero at t=4, reaching +10 at t=8]

19. A rope is towing a 100-kg rock, from rest, with a force that varies with time, as shown in the graph. At the end of eight s, the velocity of the rock is
    a) Zero
    b) -20 m/s
    c) +20 m/s
    d) +40 m/s
    e) +80 m/s

    Answer: a Difficulty: 2

20. If you pitch a baseball with twice the kinetic energy you gave it in the previous pitch, the magnitude of its momentum is
    a) the same.
    b) $\sqrt{2}$ times as much.
    c) doubled.
    d) 4 times as much.

    Answer: b Difficulty: 2

21. Car A (mass = 1000 kg) moves to the right along a level, straight road at a speed of 6 m/s. It collides directly with car B (mass = 200 kg) in a completely inelastic collision. What is the momentum after the collision if car B was initially at rest?
a) Zero
b) 6000 kg-m/s to the right.
c) 2000 kg-m/s to the right.
d) 10,000 kg-m/s to the right.
e) None of the other choices is correct.

Answer: b Difficulty: 2

22. A proton, of mass m, at rest, is struck by an alpha-particle (which consists of 2 protons and 2 neutrons) moving at velocity +v. If the collision is completely elastic, what speed will the α-particle have after the collision? (Assume the neutron's mass equals the proton's mass.)
a) Zero
b) 2/3 v
c) 3/5 v
d) 5/3 v

Answer: c Difficulty: 2

23. A bullet, of mass 20 g, traveling at 350 m/s, strikes a steel plate at an angle of 30° with the plane of the plate. It ricochets off at the same angle, at a speed of 320 m/s. What is the magnitude of the impulse that the wall gives to the bullet?
a) 0.30 N·s
b) 0.52 N·s
c) 6.7 N·s
d) 300 N·s

Answer: c Difficulty: 2

Chapter 6

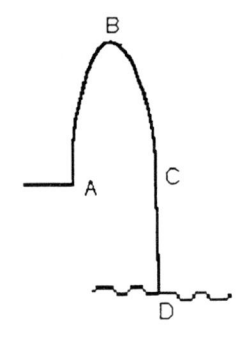

24. An Olympic diver dives off the high-diving platform. The magnitude of his momentum will be a maximum at point
   a) A.
   b) B.
   c) C.
   d) D.

   Answer: d Difficulty: 2

ESSAY

25. A mine railroad car, of mass 200 kg, rolls with negligible friction on a horizontal track with a speed of 10 m/s. A 70-kg stunt man drops straight down a distance of 4 m, and lands in the car. How fast will the car be moving after this happens?

   Answer: 7.41 m/s Difficulty: 2

MULTIPLE CHOICE

26. A 2000-kg car, traveling to the right at 30 m/s, collides with a brick wall and comes to rest in 0.2 s. The average force the car exerts on the wall is
   a) 12,000 N to the right.
   b) 300,000 N to the right.
   c) 60,000 N to the right.
   d) none of the above.

   Answer: b Difficulty: 2

ESSAY

27. A 60-kg person walks on a 100-kg log at the rate of 0.8 m/s (with respect to the log). With what speed does the log move, with respect to the shore?

   Answer: 0.3 m/s Difficulty: 2

MULTIPLE CHOICE

28. A plane, flying horizontally, releases a bomb, which explodes before hitting the ground. Neglecting air resistance, the center of mass of the bomb fragments, just after the explosion
    a) is zero.
    b) moves horizontally.
    c) moves vertically.
    d) moves along a parabolic path.

    Answer: d Difficulty: 2

29. A ball, of mass 100 g, is dropped from a height of 12 m. Its momentum when it strikes the ground is, in kg·m/s,
    a) 1.5
    b) 1.8
    c) 2.4
    d) 4.8

    Answer: a Difficulty: 2

30. A toy rocket, of mass 120 g, achieves a velocity of 40 m/s after 3 s, when fired straight up. What average thrust force does the rocket engine exert?
    a) 1.2 N
    b) 1.6 N
    c) 2.8 N
    d) 4.4 N

    Answer: b Difficulty: 2

31. A fire hose is turned on the door of a burning building in order to knock the door down. This requires a force of 1000 N. If the hose delivers 40 liters per second, what is the maximum velocity of the stream needed, assuming the water doesn't bounce back?
    a) 15 m/s
    b) 20 m/s
    c) 40 m/s
    d) 50 m/s

    Answer: b Difficulty: 2

32. A sailboat of mass m is moving with a momentum p. Which of the following represents its kinetic energy?
    a) $p^2/2m$
    b) $1/2\ mp^2$
    c) mp
    d) mp/2

    Answer: a Difficulty: 2

Chapter 6

33. Some automobiles have air bags installed on their dashboards as a safety measure. The impact of a collision causes the bag to inflate, and it then cushions a passenger's head when he is thrown forward in the car. Suppose that such a bag could lengthen the collision time between one's head and the dashboard by a factor of 10. What effect would this have on the force exerted on the head?
    a) The force would not be reduced, but the energy transferred to the head would be, thereby minimizing the damage done.
    b) The force would be reduced by a factor of about 3.
    c) The force would be reduced by a factor of about 10.
    d) The force would be reduced by a factor of about 100.
    e) The force would be reduced by a factor of about 1000.

    Answer: c Difficulty: 2

34. A small bomb, of mass 1.0 kg, is moving toward the North with a velocity of 4 m/s. It explodes into three fragments: a 5-kg fragment moving west with a speed of 8 m/s; a 4-kg fragment moving east with a speed of 10 m/s; and a third fragment with a mass of 1 kg. What is the velocity of the third fragment? (Neglect air friction.)
    a) Zero
    b) 40 m/s north
    c) 40 m/s south
    d) None of the above

    Answer: b Difficulty: 2

ESSAY

35. An arrow of mass 20 g, is shot horizontally into a bale of hay, striking the hay with a constant velocity of 60 m/s. It penetrates a depth of 20 cm before stopping. What is the average stopping force acting on the arrow?

    Answer: 180 N Difficulty: 2

36. A machine gun, of mass 35 kg, fires 50-gram bullets, with a muzzle velocity of 750 m/s, at the rate of 300 rounds per minute. What is the average force exerted on the machine gun mount?

    Answer: 188 N Difficulty: 2

MULTIPLE CHOICE

37. A golf ball traveling 3 m/s to the right collides in a head-on collision with a stationary bowling ball in a friction-free environment. If the collision is almost perfectly elastic, the speed of the golf ball immediately after the collision is
a) slightly less than 3 m/s.
b) slightly greater than 3 m/s.
c) equal to 3 m/s.
d) much less than 3 m/s.

Answer: a   Difficulty: 2

38. A 100-kg football linebacker moving at 2 m/s tackles head-on an 80-kg halfback running 3 m/s. Neglecting the effects due to digging in of cleats,
a) the linebacker will drive the halfback backward.
b) the halfback will drive the linebacker backward.
c) neither player will drive the other backward.
d) this is a simple example of an elastic collision.

Answer: b   Difficulty: 2

39. A small car meshes with a large truck in a head-on collision. Which of the following statements concerning the magnitude of the average collision force is correct?
a) The truck experiences the greater average force.
b) The small car experiences the greater average force.
c) The small car and the truck experience the same average force.
d) It is impossible to tell since the masses and velocities are not given.

Answer: c   Difficulty: 2

40. Water runs out of a horizontal drainpipe at the rate of 120 liters per minute. It falls 3.2 m to the ground. Assuming the water doesn't splash up, what average force does it exert on the ground?
a) 6.20 N
b) 12.0 N
c) 15.8 N
d) 19.6 N

Answer: c   Difficulty: 2

Chapter 6

41. A 2-kg mass moving to the east at a speed of 4 m/s collides head-on in an inelastic collision with a stationary 2-kg mass. How much kinetic energy is lost during this collision?
   a) 16 J
   b) 4 J
   c) 8 J
   d) Zero

   Answer: c Difficulty: 2

42. A Ping-Pong ball moving east at a speed of 4 m/s, collides with a stationary bowling ball. The Ping-Pong ball bounces back to the west, and the bowling ball moves very slowly to the east. Which object experiences the greater magnitude impulse during the collision?
   a) Neither; both experienced the same magnitude impulse.
   b) The Ping-Pong ball
   c) The bowling ball
   d) It's impossible to tell since the velocities after the collision are unknown.

   Answer: a Difficulty: 2

43. Two equal mass balls (one blue and the other gold) are dropped from the same height, and rebound off the floor. The blue ball rebounds to a higher position. Which ball is subjected to the greater magnitude impulse during its collision with the floor?
   a) It's impossible to tell since the time intervals and forces are unknown.
   b) Both balls were subjected to the same magnitude impulse.
   c) The gold ball
   d) The blue ball

   Answer: d Difficulty: 2

ESSAY

44. A 50-gram ball moving +10 m/s collides head-on with a stationary ball of mass 100 g. The collision is elastic. What is the speed of each ball immediately after the collision?

   Answer: -3.3 m/s and +6.7 m/s Difficulty: 2

MULTIPLE CHOICE

45. A car of mass M, traveling with a velocity V, strikes a parked station wagon, whose mass is 2M. The bumpers lock together in this head-on inelastic collision. What fraction of the initial kinetic energy is lost in this collision?
    a) 1/2
    b) 1/3
    c) 1/4
    d) 2/3

    Answer: d Difficulty: 2

ESSAY

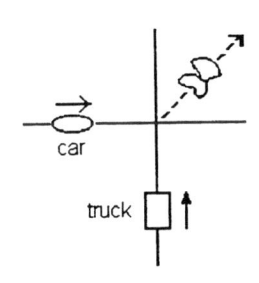

46. A car (mass = 1500 kg) and a small truck (mass = 2000 kg) collide at right angles at an icy intersection. The car was traveling east at 20 m/s and the truck was traveling north at 20 m/s when the collision took place. What is the speed of the combined wreck, assuming a completely inelastic collision.

    Answer: 14.3 m/s Difficulty: 2

47. Three masses are positioned as follows: 2 kg at (0, 0), 2 kg at (2, 0), and 4 kg at (2,1). Determine the coordinates of the center of mass.

    Answer: (1.5, 0.5) Difficulty: 2

MULTIPLE CHOICE

48. Two cars collide head-on on a level friction-free road. The collision was completely inelastic and both cars quickly came to rest during the collision. What is true about the velocity of this system's center of mass?
    a) It was always zero.
    b) It was never zero.
    c) It was not initially zero, but ended up zero.

    Answer: a Difficulty: 2

Chapter 6

49. Consider two unequal masses, M and m. Which of the following statements is <u>false</u>?
   a) The center of mass lies on the line joining the centers of each mass.
   b) The center of mass is closer to the larger mass.
   c) It is possible for the center of mass to lie within one of the objects.
   d) If a uniform rod of mass m were to join the two masses, this would not alter the position of the center of mass of the system without the rod present.

   Answer: d Difficulty: 2

50. A 3-kg mass is positioned at (0, 8), and a 1-kg mass is positioned at (12, 0). What are the coordinates of a 4-kg mass which will result in the center of mass of the system of three masses being located at the origin, (0, 0)?
   a) (-3, -6)
   b) (-12, -8)
   c) (3, 6)
   d) (-6, -3)

   Answer: a Difficulty: 2

51. The center of mass and the center of gravity coincide
   a) at all times.
   b) only if the object is symmetrical.
   c) if the object is in a uniform gravitational field.
   d) only if the object is symmetrical and it is in a uniform gravitational field.

   Answer: c Difficulty: 2

52. Two astronauts, of masses 60 kg and 80 kg, are initially at rest in outer space. They push each other apart. What is their separation after the lighter astronaut has moved 12 m?
   a) 15 m
   b) 18 m
   c) 21 m
   d) 24 m

   Answer: c Difficulty: 2

53. A right isosceles triangle is cut out from a piece of sheet metal. The two equal sides are 1 m long each. How far is the center of gravity from the right angle (90°) vertex of the triangle?
   a) 0.24 m
   b) 0.35 m
   c) 0.47 m
   d) 0.71 m

   Answer: c Difficulty: 2

ESSAY

54. What is the difference between "center of gravity" (cg) and "center of mass" (cm)? When are they the same?

Answer:
The cm is at the average position of the masses and does not depend upon gravity. The cg is the average position of weights with each mass being weighted by the strength of gravity g. The cm and the cg are identical when g is the same for all the masses.
Difficulty: 2

FILL-IN-THE-BLANK

55. A 0.17 kg baseball is thrown with a speed of 38.m/s and it is hit straight back to the pitcher with a speed of 62.m/s. What is the magnitude of the **IMPULSE** exerted upon the ball by the bat?

Answer: 17.N-s   Difficulty: 2

MULTIPLE CHOICE

56. A 2.kg ball of putty moving at 1.m/s collides and sticks to a 3.kg ball initially at rest. The putty and ball then move with a momentum of:
    a) 0 N-s
    b) 1 N-s
    c) 2 N-s
    d) 3 N-s
    e) 4 N-s
    f) 5 N-s
    g) more than 5 N-s

Answer: b   Difficulty: 2

TRUE/FALSE

57. Automobile air bags reduce the IMPULSE upon a person during a collision.

Answer: F   Difficulty: 1

MULTIPLE CHOICE

58. A car hits another and the two bumpers lock together during the collision. Is this an elastic or inelastic collision?
    a) elastic
    b) inelastic
    c) insufficient information (it could be either)

Answer: b   Difficulty: 2

Chapter 6

59. Suppose a bug spatters upom your windshield as you drive down the road. Which sustains the greater change of momentum?
    a) the car
    b) the bug
    c) both the same
    d) insufficient information

    Answer: c Difficulty: 1

60. Which of the following has units of momentum?
    a) kg-s/m
    b) kg-m^2/s
    c) kg-m^2/s^2
    d) J-s/m
    e) N-m

    Answer: d Difficulty: 1

FILL-IN-THE-BLANK

61. Three masses are positioned at the following coordinates: 3.0 kg at (3,2), and 4.0 kg at (0,-1), and 5.0 kg at (5,-7). What are the coordinates of the center of mass of the system?

    Answer: (2.8 , -2.8) Difficulty: 2

62. An 80.kg man is skating northward and happens to collide with a 20.kg boy who is ice skating toward the east. Immediately after the collision, the man and boy are seen to be moving together at 2.5 m/s in a direction 60 degrees north of east. How fast was the boy moving before the collision?

    Answer:
    6.3 m/s

    Difficulty: 2

63. An empty coal-car (mass 20,000.kg) of a train coasts along at 10.m/s. An unfortunate 3000.kg elephant falls from a bridge and drops vertically into the car. Determine the speed of the car immediately after the elephant is added to its contents.

    Answer: 8.7 m/s Difficulty: 2

MULTIPLE CHOICE

64. Two cars suffer a head-on collision. What is conserved in the collision?
   a) kinetic energy
   b) mechanical energy and momentum
   c) momentum
   d) none of these

   Answer: c Difficulty: 2

TRUE/FALSE

65. A 1.0 kg cheese-ball vertically falls and strikes the floor with a speed of 20.m/s and it bounces straight up with a speed of 15.m/s. The magnitude of its change of momentum is 5. kg-m/s.

   Answer: F Difficulty: 2

ESSAY

66. Discuss similarities and differences between KINETIC ENERGY and MOMENTUM.

   Answer:
   K is a scalar whereas P is a vector. Both depend upon mass and speed, however K depends upon the square of the speed but P is simply proportional to velocity. Momentum is always conserved for an isolated system but K might change if it changes to/from other energy forms.
   Difficulty: 2

FILL-IN-THE-BLANK

67. Momentum of a system isn't always constant. What condition assures that momentum is conserved for a system?

   Answer:
   Momentum is conserved for an "isolated" system; i.e., if the net external force acting on the system is zero.
   Difficulty: 2

68. Jennifer hits a stationary 200.gram ball and it leaves the racket at 40.m/s. If time lapse photography shows that the ball was in contact with the racket for 40.ms.
   a) What average force was exerted on the racket?
   b) What is the ratio of this force to the weight of the ball?

   Answer: a) 0.20 kN   b) 1.0×10^2 Difficulty: 2

Chapter 6

TRUE/FALSE

69. In an inelastic collision, the initial kinetic energy is less than the final kinetic energy.

    Answer: F Difficulty: 2

FILL-IN-THE-BLANK

70. In space, a 4.0 kg metal ball moving 30.m/s has a head on collision with a stationary similar second ball. After the elastic collision, what are the velocities of the balls?

    Answer:
    They exchange velocities: first becomes zero, and the 2nd acquired 30.m/s in the direction the 1st was moving.
    Difficulty: 2

71. In space, a 4.0 kg metal ball moving 30.m/s has a head on collision with a stationary 1.0 kg second ball. After the elastic collision, what are the velocities of the balls?

    Answer:
    The first becomes 18.m/s, and the 2nd acquires 48.m/s; both in the direction the 1st was moving.
    Difficulty: 2

72. A locomotive elastically collides head on with a stationary ball. In terms of the locomotives initial speed, how fast is the ball moving after the collision?

    Answer: the ball acquires TWICE the locomotive's velocity Difficulty: 2

MULTIPLE CHOICE

73. A 30.kg child stands at one end of a floating 20.kg canoe that is 10.ft long. The child walks to the other end of the canoe. How far does the canoe move in the water assuming water friction is negligible?
    a) 2.ft
    b) 4.ft
    c) 6.ft
    d) 8.ft
    e) 10.ft
    f) none of the above

    Answer: c Difficulty: 2

ESSAY

74. In space there is nothing for the rocket to "push against" so how does it accelerate?

    Answer:
    In terms of forces, the rocket pushes the exhaust gases back and they in turn (action-reaction) push the rocket forward. In terms of momentum, in order to conserve momentum since the exhaust gases acquire negative momentum (moving backwards) the rocket acquires positive momentum (increasing velocity forward).

    Difficulty: 2

Chapter 7

MATCHING

Match the unit to the physical quantity.

a) rad/s^2
b) s^2/m^3
c) Hz
d) rad/s
e) J
f) seconds
g) N·m^2/kg^2
h) m/s
i) radians

1. angular acceleration

    Answer: a) rad/s^2 Difficulty: 2

2. Kin Kepler's Third Law

    Answer: b) s^2/m^3 Difficulty: 2

3. frequency

    Answer: c) Hz Difficulty: 2

4. angular velocity

    Answer: d) rad/s Difficulty: 2

5. gravitational potential energy

    Answer: e) J Difficulty: 2

6. period

    Answer: f) seconds Difficulty: 2

7. universal Gravitational constant

    Answer: g) N·m^2/kg^2 Difficulty: 2

8. escape velocity

    Answer: h) m/s Difficulty: 2

9. angular distance

    Answer: i) radians Difficulty: 2

ESSAY

10. What is the arc length subtended by an angle of 30° on a circle of radius 10 cm?

    Answer: 5.24 cm Difficulty: 2

11. A triangle has angles of 1.10 radians and 0.25 radians. What is the third angle, expressed in radians and in degrees?

    Answer: 1.79 rad, and 103° Difficulty: 2

12. A bicycle wheel has an outside diameter of 66 cm. How far does a point on the rim travel when the wheel rotates through an angle of 70°?

    Answer: 40.3 cm Difficulty: 2

MULTIPLE CHOICE

13. A flywheel, of radius 1 m, is spinning with a constant angular velocity of 2 rad/s. What is the centripetal acceleration of a point on the wheel's rim?
    a) 1 m/s^2
    b) 2 m/s^2
    c) 4 m/s^2
    d) None of the above

    Answer: c Difficulty: 2

TRUE/FALSE

14. A car, driven around a circle with constant speed, must have zero acceleration.

    Answer: F Difficulty: 2

ESSAY

15. A tooth polisher on a dentist's drill reaches a frequency of 1800 rpm in 8 s. What is its average angular acceleration?

    Answer: 23.6 rad/s^2 Difficulty: 2

Chapter 7

MULTIPLE CHOICE

16. A stone, of mass m, is attached to a strong string and whirled in a vertical circle of radius r. At the exact top of the path the tension in the string is 3 times the stone's weight. The stone's speed at this point is given by
    a) $2\sqrt{gr}$
    b) $\sqrt{gr}$
    c) $4gr$
    d) $\sqrt{2gr}$

    Answer: a Difficulty: 2

ESSAY

17. What is the angular speed of an electric motor that rotates at 1800 rpm?

    Answer: 188 rad/s Difficulty: 2

MULTIPLE CHOICE

18. A fan blade, whose diameter is 1 m, is turning with an angular velocity of 2 rad/s. What is the tangential velocity of a point on the tip of the blade?
    a) 2 m/s
    b) 0.5 m/s
    c) 1 m/s
    d) None of the above

    Answer: c Difficulty: 2

ESSAY

19. What is the angular speed of a record that rotates a 33-1/3 rpm?

    Answer: 3.49 rad/s Difficulty: 2

20. The flywheel of a machine is rotating at 126 rad/s. Through what angle will the wheel be displaced from its original position after 6 s?

    Answer: 116° Difficulty: 2

Testbank

MULTIPLE CHOICE

21. The maximum speed around a level curve is 30 mi/h. What is the maximum speed around a curve with twice the radius? (Assume all other factors remain unchanged.)
    a) 42.4 mi/h
    b) 45 mi/h
    c) 60 mi/h
    d) 120 mi/h
    e) 21.1 mi/h

    Answer: a Difficulty: 2

ESSAY

22. What is the angular speed of a point on the earth's surface at 60° north latitude? What is the linear speed of such a point?

    Answer: 7.27×10^{-5} rad/s; 0.232 km/s Difficulty: 2

MULTIPLE CHOICE

23. A race car, traveling at a constant speed of 50 m/s, drives around a circular track of radius 250 m. What angular acceleration does it experience?
    a) zero
    b) 10 rad/s^2
    c) 625 rad/s^2
    d) 10^4 rad/s^2

    Answer: a Difficulty: 2

24. What is the centripetal acceleration of a point on the perimeter of a bicycle wheel of diameter 70 cm when the bike is moving 8 m/s?
    a) 91 m/s^2
    b) 183 m/s^2
    c) 206 m/s^2
    d) 266 m/s^2

    Answer: b Difficulty: 2

25. Two horizontal curves on a bobsled run are banked at the same angle, but one has twice the radius of the other. The safe speed (no friction needed to stay on the run) for the smaller radius curve is V. Therefore, the safe speed on the larger radius curve is
    a) approximately 0.707 V.
    b) 2 V.
    c) approximately 1.41 V.
    d) 0.5 V.

    Answer: c Difficulty: 2

Chapter 7

26. The banking angle in a turn on the Olympic bobsled track is not constant, but increases upward from the horizontal. Coming around a turn, the bobsled team will intentionally "climb the wall," then go lower coming out of the turn. Why do they do this?
    a) To give the team better control, because they are able to see ahead of the turn.
    b) To prevent the bobsled from turning over.
    c) To take the turn at a faster speed.
    d) To reduce the g-force on them.

    Answer: c Difficulty: 2

27. A horizontal curve on a bobsled run is banked at a 45° angle. When a bobsled rounds this curve at the curve's safe speed (no friction needed to stay on the run), what is its centripetal acceleration?
    a) 1.0 g
    b) 2.0 g
    c) 0.5 g
    d) None of the above

    Answer: a Difficulty: 2

ESSAY

28. The diameter of the sun is 13.9×10^8 m, and the distance from the earth to the sun is 1.5×10^{11} m. How long does it take for the sun to disappear after it touches the horizon?

    Answer: 2.12 min Difficulty: 2

TRUE/FALSE

29. You can ride your bicycle in a circle while your wheels are perpendicular to the ground.

    Answer: F Difficulty: 2

MULTIPLE CHOICE

30. Does the centripetal force acting on an object do work on the object?
    a) Yes, since a force acts and the object moves, and work is force times distance.
    b) Yes, since it takes energy to turn an object.
    c) No, because the object has constant speed.
    d) No, because the force and the displacement of the object are perpendicular.

    Answer: d Difficulty: 2

Testbank

ESSAY

31. What is the centripetal acceleration of a person at the equator, expressed in a multiple of g?

   Answer: 0.034 g   Difficulty: 2

MULTIPLE CHOICE

32. A car traveling 20 m/s rounds an 80-m radius horizontal curve with the tires on the verge of slipping. How fast can this car round a second curve of radius 320 m? (Assume the same coefficient of friction between the car's tires and each road surface.)
   a) 40 m/s
   b) 30 m/s
   c) 80 m/s
   d) None of the above

   Answer: a   Difficulty: 2

33. A roller coaster car is on a track that forms a circular loop in the vertical plane. If the car is to just maintain contact with track at the top of the loop, what is the minimum value for its centripetal acceleration at this point?
   a) g downward
   b) 0.5 g downward
   c) g upward
   d) 2 g upward

   Answer: a   Difficulty: 2

34. A roller coaster car (mass = M) is on a track that forms a circular loop (radius = r) in the vertical plane. If the car is to just maintain contact with the track at the top of the loop, what is the minimum value for its speed at that point?
   a) 2 Mrg
   b) $\sqrt{2rg}$
   c) $\sqrt{rg}$
   d) None of the above

   Answer: c   Difficulty: 2

35. The speed needed at the bottom of a loop-the-loop track so that a car can coast to the top, with sufficient speed to stay on the track, depends on the mass of the car. (Neglect the effects of friction.)
   a) Always true.
   b) Never true.
   c) Sometimes true, since kinetic energy is a function of mass.
   d) Sometimes true, since potential energy is a function of mass.

   Answer: b   Difficulty: 2

Chapter 7

TRUE/FALSE

36. The escape velocity is twice the orbital velocity.

    Answer: F Difficulty: 2

MULTIPLE CHOICE

37. Who was the first person to realize that the planets move in elliptical paths around the sun?
    a) Kepler
    b) Brahe
    c) Einstein
    d) Copernicus

    Answer: a Difficulty: 2

ESSAY

38. What is the gravitational force on a 70-kg person, due to the moon? The mass of the moon is 7.36×10^{22} kg and the distance to the moon is 3.82×10^{8} m.

    Answer: 0.0024 N Difficulty: 2

TRUE/FALSE

39. The net force on an object in orbit is zero.

    Answer: F Difficulty: 2

MULTIPLE CHOICE

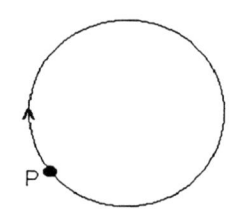

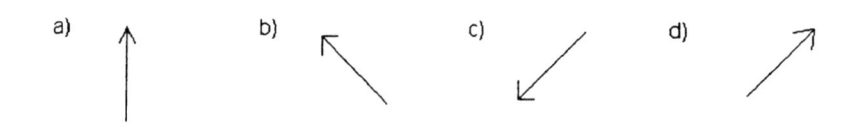

40. A planet revolves clockwise around a star, with constant speed. The direction of its acceleration at point P is

    Answer: d Difficulty: 2

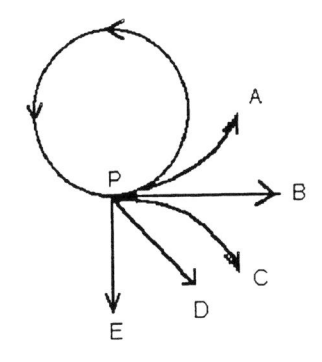

41. A girl attaches a rock to a string, which she then swings counter-clockwise in a horizontal circle. The string breaks at point P on the sketch, which shows a bird's-eye view (i.e., as seen from above). What path will the rock follow?
   a) A
   b) B
   c) C
   d) D
   e) E

   Answer: b Difficulty: 2

ESSAY

42. For a spacecraft going from the earth toward the sun, at what point will the gravitational forces due to the sun and the earth cancel?
    earth's mass: $m_e = 5.98 \times 10^{24}$ kg
    sun's mass: $m_s = 1.99 \times 10^{30}$ kg
    earth-sun distance: $r = 1.50 \times 10^{11}$ m

    Answer: 2.60×10^8 m from earth Difficulty: 2

MULTIPLE CHOICE

43. A satellite encircles Mars at a distance above its surface equal to 3 times the radius of Mars. The acceleration of gravity of the satellite, as compared to the acceleration of gravity on the surface of Mars, is
    a) zero.
    b) the same.
    c) one-third as much.
    d) one-ninth as much.
    e) one-sixteenth as much.

    Answer: e Difficulty: 2

Chapter 7

ESSAY

44. The maximum acceleration a pilot can stand is about 7 g. What is the minimum radius of curvature that a jet plane's pilot, pulling out of a vertical dive, can tolerate at a speed of 250 m/s?

    Answer: 1060 m Difficulty: 2

MULTIPLE CHOICE

45. The rotating space station you're standing in simulates earth's gravity when you are 500 ft from its center. What must its angular speed be?
    a) 0.064 rad/s
    b) 0.14 rad/s
    c) 0.25 rad/s
    d) 126 rad/s

    Answer: c Difficulty: 2

ESSAY

46. By how many newtons does the weight of a 100-kg person change when he goes from sea level to an altitude of 5000 m?

    Answer: 1.54 N Difficulty: 2

MULTIPLE CHOICE

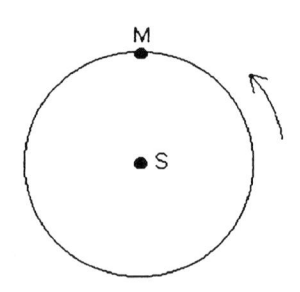

47. Marilyn (M) and her sister Sheila (S) are riding on a merry-go-round, as shown (bird's-eye view).
    a) They have the same speed, but their angular velocity is different.
    b) They have different speeds, and different angular velocities.
    c) They have different speeds, but the same angular velocity.
    d) They have the same speed and the same angular velocity.

    Answer: c Difficulty: 2

ESSAY

48. Our second closest star, Alpha Centauri, is 4 ly away. On a particular night, it is located 37° up from the horizon, in a vertical plane. What are its x and y coordinates in this plane?

    Answer: x = 3.2 ly, y = 2.4 ly   Difficulty: 2

49. A satellite is in a low circular orbit about the earth (i.e, it just skims the surface of the earth). How long does it take to make one revolution around the earth?

    Answer: 86.8 min   Difficulty: 2

MULTIPLE CHOICE

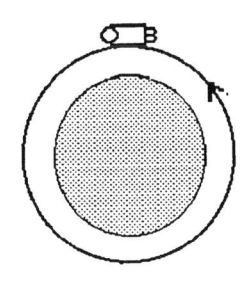

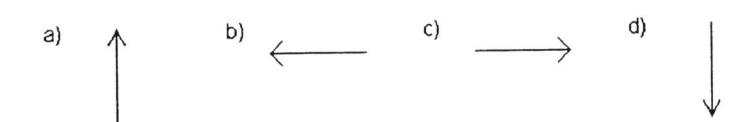

50. Your starship, the TOP-QUARK, is in a counter-clockwise "parking orbit" around the planet VULCAN, as shown, in order to achieve ascape velocity, the direction your retro-rockets should be facing is

    Answer: c   Difficulty: 2

51. A pulsar (a rotating neutron star) emits pulses at a frequency of 0.4 kHz. The period of its rotation is
    a) 2.5 ms
    b) 2.5 s
    c) 0.025 s
    d) 25 ms

    Answer: a   Difficulty: 2

Chapter 7

52. A pilot executes a vertical dive. Just before the plane starts to come up from the bottom of the dive, the force on him is
a) less than g, and pointing up.
b) less than g, and pointing down.
c) more than g, and pointing up.
d) more than g, and pointing down.

Answer: c Difficulty: 2

53. A jet plane flying 600 m/s experiences an acceleration of 4g when pulling out of the dive. What is the radius of curvature of the loop in which the plane is flying? (For this problem use $g = 10$ m/s^2.)
a) 640 m
b) 1200 m
c) 7000 m
d) 9000 m

Answer: d Difficulty: 2

ESSAY

54. A pilot makes an outside vertical loop of radius 320 m. At the top of his loop he is pushing down on his seat with only one-half of his normal weight. How fast is he going?

Answer: 39.6 m/s Difficulty: 2

MULTIPLE CHOICE

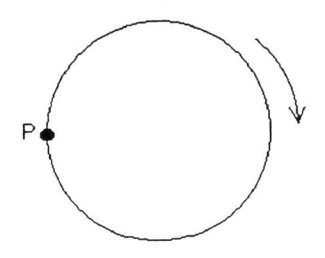

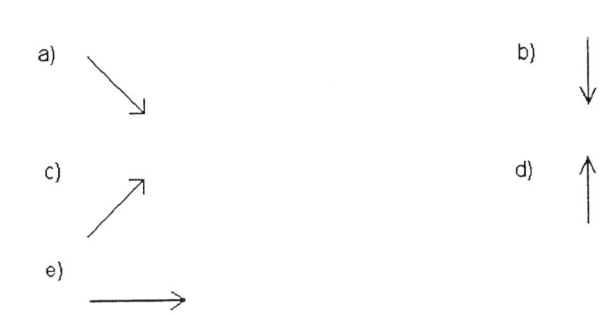

a) ↘
b) ↓
c) ↗
d) ↑
e) →

55. The record playing on the turntable of your stereo is rotating clockwise (as seen from above). After turning it off, your turntable is slowing down, but hasn't stopped yet. The direction of the acceleration of point P (at the left) is

    Answer: a Difficulty: 2

56. A spaceship is traveling to the moon. At what point is it beyond the pull of earth's gravity?
    a) When it gets above the atmosphere
    b) When it is half-way there
    c) When it is closer to the moon than it is to earth
    d) It is never beyond the pull of earth's gravity.

    Answer: d Difficulty: 2

57. Satellite A has twice the mass of satellite T, and rotates in the same orbit.
    a) The speed of T is twice the speed of A.
    b) The speed of T is half the speed of A.
    c) The speed of T is one-fourth the speed of A.
    d) The speed of T is equal to the speed of A.

    Answer: d Difficulty: 2

Chapter 7

58. A moon is in orbit around its mother planet. Which of the following statements is always true about its kinetic energy (K), and its gravitational potential energy (U)?
a) K is positive, and U is positive.
b) K is negative, and U is positive.
c) K is positive, and U is negative.
d) K is negative, and U is negative.

Answer: c Difficulty: 2

59. Two asteroids, of the same mass, have a gravitational potential energy U. If a third asteroid, of the same mass, moves toward them, to a position such that the distances between the three asteroids are the same, the gravitational potential energy of the system will have changed by
a) U
b) 2U
c) 3U
d) 4U

Answer: b Difficulty: 2

60. Consider a small satellite moving in a circular orbit (radius r) about a spherical planet (mass M). Which expression gives this satellite's orbital velocity?
a) $v = GM/r$
b) $\sqrt{GM/r}$
c) $v = \sqrt{GM/r^2}$
d) $\sqrt{Gr}$

Answer: b Difficulty: 2

61. A spherically symmetric planet has four times the earth's mass and twice its radius. If a jar of peanut butter weighs 12 N on the surface of the earth, how much would it weigh on the surface of this planet?
a) 6 N
b) 12 N
c) 24 N
d) None of the above

Answer: b Difficulty: 2

ESSAY

62. An earth satellite is in circular orbit 230 km above the surface of the earth. It is observed to have a period of 89 min. From this information, estimate the mass of the earth.

Answer: 6.0×10^{24} kg Difficulty: 2

63. Europa, a moon of Jupiter, has an orbital diameter of 1.34×10^9 m, and a period of 3.55 days. What is the mass of Jupiter?

   Answer: 1.89×10^{27} kg Difficulty: 2

MULTIPLE CHOICE

64. An astronaut goes out for a "space-walk" at a distance above the earth equal to the radius of the earth. Her acceleration will be
    a) zero
    b) g
    c) 1/2 g
    d) 1/4 g

   Answer: d Difficulty: 2

TRUE/FALSE

65. The constant K in Kepler's Third Law is the same for all the planets in our solar system, but is different for planetary systems orbiting other stars.

   Answer: T Difficulty: 2

MULTIPLE CHOICE

66. A planet is discovered to orbit around a star in the galaxy Andromeda, with the same orbital diameter as the earth around our sun. If that star has 4 times the mass of our Sun, what will the period of revolution of that new planet be, compared to the earth's orbital period?
    a) One-fourth as much
    b) One-half as much
    c) Twice as much
    d) Four times as much

   Answer: b Difficulty: 2

67. The speed of Halley's comet, while traveling in its elliptical orbit around the sun,
    a) is constant.
    b) increases as it nears the sun.
    c) decreases as it nears the sun.
    d) is zero at two points in the orbit.

   Answer: b Difficulty: 2

Chapter 7

68. The following statements refer to man-made, artificial satellites in orbit around earth. Which is an accurate statement?
   a) It is possible to have a satellite traveling at either a high speed or at a low speed in a given circular orbit.
   b) Only circular orbits (and not elliptical ones) are possible for artificial satellites.
   c) A satellite in a large diameter circular orbit will always have a longer period of revolution about the earth than will a satellite in a smaller circular orbit.
   d) The velocity required to keep a satellite in a given orbit depends on the mass of the satellite.
   e) The period of revolution of a satellite moving about the earth is independent of the size of the orbit it travels.

   Answer: c   Difficulty: 2

ESSAY

69. A future use of space stations may be to provide hospitals for severely burned persons. It is very painful for a badly burned person on earth to lie in bed. In a space station, the effect of gravity can be reduced or eliminated, At what frequency, in rpm, would a doughnut-shaped hospital of 200 m radius have to rotate, if persons on the outer perimeter are to experience 1/10 the gravity effect of earth?

   Answer: 0.011 Hz   Difficulty: 2

70. A coin of mass m rests on a turntable a distance r from the axis of rotation. The turntable rotates with frequency f. Derive an expression for the minimum coefficient of friction between the turntable and the coin if the coin is not to slip.

   Answer: $(4\pi^2 r f^2)/g$ Difficulty: 2

71. Why do the two different expressions for gravitational potential energy ($+mgh$; $-GmM/r^2$) have different signs?

   Answer:
   The reference level for zero grav. pot. energy is different. The mgh assumes the zero level is where the height h is zero; but $-GmM/r^2$ infers the zero level is at infinity. As we move away from the Earth, both expressions must yield greater pot. energy. Indeed mgh increases as h increases, and $-GmM/r^2$ gets less negative (more positive) as r increases too.

   Difficulty: 2

MULTIPLE CHOICE

72. A space station of diameter 40. meters is turning about its axis at a constant rate. What is the period of revolution of the space station if the outer rim experiences an acceleration of 2.5 m/s^2 ?
    a) 13. s
    b) 10. s
    c) 11. s
    d) 14. s
    e) 18. s

    Answer: e   Difficulty: 2

FILL-IN-THE-BLANK

73. A satellite orbits the Earth once every 6.0 hours in a circle.

    a) What is its angular velocity (in radians/s)?

    b) What is the radius of the orbit if it is a 2000. Kg satellite?

    c) What is the acceleration of the satellite (direction too)?

    Answer: a) 2.9 x 10^{-4} rad/s   b) 1.7 x 10^7 m   c) 1.4 m/s^2

    Difficulty: 2

74. Consider a 2.4 kilogram feather (Big Bird) on the planet Mars. [Earth mass=6.0x10^{24} kg = 9.3 x Mars mass; Earth radius= 6380. km =1.89xMars radius]

    a) What does the feather weigh on Mars? _____

    b) What would the same feather weight on Earth? _____

    c) What is the acceleration of gravity on Mars? _____

    Answer: a) 9.1 N   b) 24. N   c) 3.8 m/s^2 Difficulty: 2

Chapter 7

MULTIPLE CHOICE

75. The speed for a "low" circular orbit about the earth is about:
    a) 2.1 km/s
    b) 7.6 km/s
    c) 17. km/s
    d) 17,000. km/s
    e) 25,000. km/s
    f) 2200. m/s

    Answer: b Difficulty: 2

76. During the period of one day, the number of high tides on the east or west coast is usually:
    a) one
    b) two
    c) three
    d) four
    e) none of these

    Answer: b Difficulty: 2

TRUE/FALSE

77. Kepler's discovery that $T^2=Kr^3$ applies only to circular orbits.

    Answer: T Difficulty: 1

FILL-IN-THE-BLANK

78. The weight of an object on the Moon is what fraction of the weight of the same object on Earth? _____

    Answer: 1/6 Difficulty: 1

MULTIPLE CHOICE

79. The height of the tides is influenced by:
    a) the moon
    b) the sun
    c) the uneven surface of the Earth
    d) all of the above
    e) none of the above

    Answer: d Difficulty: 2

80. If the Earth had four times its present mass, what would be its new period of revolution around the Sun, compared to its present orbital period?
    a) One-fourth as much
    b) Four times as much
    c) Two times as much
    d) One-half as much
    e) the same

    Answer: e Difficulty: 2

FILL-IN-THE-BLANK

81. A 63.kg astronaut walks upon the surface of the planet Krypton which has 100.times the mass of the Earth and 100.times the diameter of the Earth. What does she weigh on Krypton?

    Answer: $63. \times 9.8 \times 100/100^2 =$ 6.2 Newtons Difficulty: 3

MULTIPLE CHOICE

82. If the newly discovered planet of Krypton has 100.times the mass of the Earth and 100.times the diameter of the Earth, what would be the period of a "low" orbiting satellite around Krypton?
    a) 1500. hours
    b) 150. hours
    c) 15. hours
    d) 1.5 hours

    Answer: c Difficulty: 2

ESSAY

83. Imagine yourself in a car traveling around a sharp curve. The "centripetal" force is inward yet why do you slide outward and not inward?

    Answer:
    In the absence of an inward force, you want to continue in a straight line which would cause you to move "outward". The frictional force of the seat or the push of the side of the car pushes you "inward" and supplies the centripetal force to keep you on your circular path.

    Difficulty: 2

Chapter 7

TRUE/FALSE

84. James moves in a circular path (radius 15.m) with a speed of 60.m/s. His ANGULAR VELOCITY is $120/\pi$ revolutions per minute.

    Answer: T   Difficulty: 2

ESSAY

85. How was Newton able to calculate the period of revolution of the Moon using his gravitation theory, even though he didn't know the value of G nor the masses of the Earth and Moon?

    Answer:
    From $GmM/r^2 = mv^2/r$ where m,M are lunar and Earth masses and r distance from Earth to Moon; he got period $T = 2\pi\, r^{3/2}/(GM)^{1/2}$
    But he knew $g = GM/R^2$ where R is Earth radius, so
    $T = 2\pi\, r^{3/2}/(gR^2)^{1/2}$ and he knew g, r, and R
    Difficulty: 3

86. Show that the change in gravitational potential energy $U_2 - U_1 = -GmM(1/R_2 - 1/R_1)$ becomes mgh when h<<r.

    Answer:
    r=R+h where R=Earth radius. $U_2 - U_1 = -gmM(1/R_2 - 1/R_1) = -gmM(R_1 - R_2)/(R_1 R_2)$
    $= gmM(h)/(R_1 R_2)$; R_2=approx R_1 if h<<R so $U_2 - U_1$=mgh where $g = GM/R^2$
    Difficulty: 3

FILL-IN-THE-BLANK

87. Visiting the local carnival, Suzy notices her friends on a "ride" in which they were experiencing 3.0 revolutions every 4 seconds on a circle of 12.meter radius.
    a) Her friends were experiencing what centripetal acceleration?
    b) How many "g's" is this?

    Answer: a) 22.m/s^2 b) 2.3 g's   Difficulty: 2

88. A turntable was revolving at 33 1/3 rpm. It was shut off and uniformly slowed down and stopped in 5.5 seconds.
    a) What was the angular acceleration?
    b) Through how many revolutions did it turn?

    Answer: a) -0.10 rev/s^2 = -0.63 radians/s^2 b) 1.5 revolutions

    Difficulty: 2

MULTIPLE CHOICE

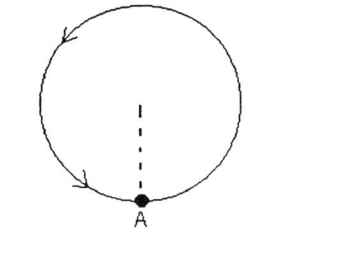

(A) ↓   (B) ↘   (C) →   (D) ↗

89. James is twirling a rock around a horizontal circle on the end of a string. See the accompanying figure which is a top view. Suppose the string is cut when the rock reaches point A. The rock then moves off in what direction?

    Answer: c Difficulty: 2

TRUE/FALSE

90. Newton never knew the value of "his" universal gravitational constant G.

    Answer: T Difficulty: 1

FILL-IN-THE-BLANK

91. As you know, the Earth has an orbital period of one year at an orbital radius of one AU. If a new "minor planet" were to be found in a circular orbit with radius 13.AU, what would be its period?

    Answer: Remembering Kepler's 3rd law: $T^2 = k R^3$ so T = 47.years at R=13.AU

    Difficulty: 2

MULTIPLE CHOICE

92. How does the escape speed from Mars compare with that of the Earth? [M(Earth)=9.31 M(Mars); R(Earth)=1.88 R(Mars)]

    a) 20.2%
    b) 22.5%
    c) 44.9%
    d) 66.2%
    e) 222.%
    f) 495.%

    Answer: c Difficulty: 2

Chapter 7

FILL-IN-THE-BLANK

93. Jenny drives at a constant speed of 15. m/s around a circular drive of diameter 60. meters.
    A) Her acceleration is _____ m/s^2
    B) Her acceleration is:
        (a) tangential, (b) centripetal, (c) centrifugal.
    C) Her velocity is:
        (a) tangential, (b) centripetal, (c) centrifugal.

Answer: A) 7.5 m/s^2 B) b   C) a   Difficulty: 2

Chapter 8

MATCHING

Match the unit to the physical quantity.

a) kg·m²/s
b) kg·m²
c) N·m
d) joules
e) feet

1. angular momentum

    Answer: a) kg·m²/s Difficulty: 2

2. moment of inertia

    Answer: b) kg·m² Difficulty: 2

3. torque

    Answer: c) N·m Difficulty: 2

4. rotational kinetic energy

    Answer: d) joules Difficulty: 2

5. lever-arm

    Answer: e) feet Difficulty: 2

MULTIPLE CHOICE

6. What condition or conditions are necessary for rotational equilibrium?
    a) $\Sigma F_x = 0$
    b) $\Sigma F_X = 0$, $\Sigma F_Y = 0$, $\Sigma \tau = 0$
    c) $\Sigma \tau = 0$
    d) $\Sigma F_Y = 0$
    e) $\Sigma F_x = 0$, $\Sigma F_Y = 0$

    Answer: c Difficulty: 2

7. What condition or conditions are necessary for static equilibrium?
    a) $\Sigma F_x = 0$
    b) $\Sigma F_X = 0$, $\Sigma F_Y = 0$, $\Sigma \tau = 0$
    c) $\Sigma \tau = 0$
    d) $\Sigma F_Y = 0$
    e) $\Sigma F_x = 0$, $\Sigma F_Y = 0$

    Answer: b Difficulty: 2

Chapter 8

8. A boy and a girl are balanced on a massless seesaw. The boy has a mass of 75 kg and the girl's mass is 50 kg. If the boy sits 2 m from the pivot point on one side of the seesaw, where must the girl sit on the other side?
   a) 1.33 m
   b) 2.5 m
   c) 3 m
   d) None of the above

   Answer: c Difficulty: 2

9. A heavy boy and a lightweight girl are balanced on a massless seesaw. If they both move forward so that they are one-half their original distance from the pivot point, what will happen to the seesaw?
   a) The side the boy is sitting on will tilt downward.
   b) The side the girl is sitting on will tilt downward.
   c) Nothing, the seesaw will still be balanced.
   d) It is impossible to say without knowing the masses and the distances.

   Answer: c Difficulty: 2

10. A heavy seesaw is out of balance. A lightweight girl sits on the end that is tilted downward, and a heavy boy sits on the other side so that the seesaw now balances. If the boy and girl both move forward so that they are one-half their original distance from the pivot point, what will happen to the seesaw?
    a) The side the boy is sitting on will now tilt downward.
    b) The side the girl is sitting on will once again tilt downward.
    c) Nothing, the seesaw will still be balanced.
    d) It is impossible to say without knowing the masses and the distances.

    Answer: b Difficulty: 2

TRUE/FALSE

11. A non-zero torque is needed to produce a change in angular velocity.

    Answer: T Difficulty: 2

Testbank

MULTIPLE CHOICE

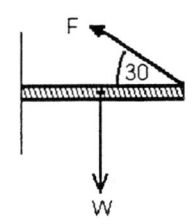

12. A force is applied to the end of a 2 ft long uniform board weighing 50 lb, in order to keep it horizontal, while it pushes against a wall at the left. If the angle the force makes with the board is 30° in the direction shown, the applied force F is
    a) 28.9 lb
    b) 50 lb
    c) 57.7 lb
    d) 100 lb

    Answer: b Difficulty: 2

13. Consider a rigid body that is rotating. Which of the following is an accurate statement?
    a) Its center of rotation is its center of gravity.
    b) All points on the body are moving with the same angular velocity.
    c) All points on the body are moving with the same linear velocity.
    d) Its center of rotation is at rest, i.e., not moving.

    Answer: b Difficulty: 2

14. A bicycle is moving 4 m/s. What is the angular speed of a wheel if its radius is 30 cm?
    a) 0.36 rad/s
    b) 1.2 rad/s
    c) 4.8 rad/s
    d) 13.3 rad/s

    Answer: d Difficulty: 2

ESSAY

15. A triatomic molecule is modeled as follows: mass m is at the origin, mass 2m is at x = a, and, mass 3m is at x = 2a. (a) What is the amount of inertia about the origin? (b) What is the moment of inertia about the center of mass?

    Answer: (a) 14 ma^2 (b) 3.3 ma^2 Difficulty: 2

16. A cylinder, of radius 8 cm, rolls 20 cm in 5 s. Through what angular displacement does the cylinder move in this time?

    Answer: 143° Difficulty: 2

Chapter 8

17. A wheel of diameter 0.7 m rolls without slipping. A point at the top of the wheel moves with a tangential speed 2 m/s. (a) At what speed is the axis of the wheel moving? (b) What is the angular speed of the wheel?

    Answer: (a) 1 m/s   (b) 2.86 rad/s   Difficulty: 2

18. Consider a bicycle wheel to be a ring of radius 30 cm and mass 1.5 kg. Neglect the mass of the axle and sprocket. If a force of 20 N is applied tangentially to a sprocket of radius 4 cm for 4 s, what linear speed does the wheel achieve, assuming it rolls without slipping?

    Answer: 7.11 m/s   Difficulty: 2

19. A cylinder of radius R and mass M rolls without slipping down a plane inclined at an angle G above horizontal. (a) What is the torque about the point of contact with the plane? (b) What is the moment of inertia about the point of contact? (c) What is the linear speed of the cylinder t s after it is released from rest?

    Answer: (a) $MgR \sin G$ (b) $3MR^2/2$ (c) $2gt \sin G$ Difficulty: 2

MULTIPLE CHOICE

20. A boy and a girl are riding on a merry-go-round that is turning. The boy is twice as far as the girl from the merry-go-round's center. If the boy and girl are of equal mass, which statement is true about the boy's moment of inertia with respect to the axis of rotation?
    a) His moment of inertia is 4 times the girl's.
    b) His moment of inertia is twice the girl's.
    c) The moment of inertia is the same for both.
    d) The boy has a greater moment of inertia, but it is impossible to say exactly how much more.

    Answer: a   Difficulty: 2

```
        | string
        |
 |hinge |
 |  rod |
wall|____|
    |
```

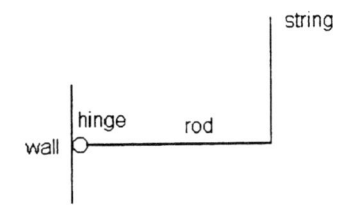

21. A uniform rod has a weight of 40 N and a length of 1 m. It is hinged to a wall (at the left end), and held in a horizontal position by a vertical massless string (at the right end). What is the magnitude of the torque exerted by the string about a horizontal axis which passes through the hinge and is perpendicular to the rod?
    a) 5 N-m
    b) 10 N-m
    c) 20 N-m
    d) 40 N-m

    Answer: c Difficulty: 2

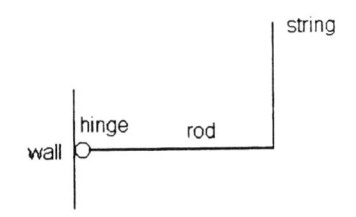

```
        | string
        |
 |hinge |
 |  rod |
wall|____|
    |
```

22. A uniform rod has a length of 2 m. It is hinged to a wall (at the left end), and held in a horizontal position by a vertical massless string (at the right end). What is the angular acceleration of the rod at the moment the string is released?
    a) 3.27 rad/s^2
    b) 14.7 rad/s^2
    c) 7.35 rad/s^2
    d) It can't be calculated without knowing the rod's mass.

    Answer: c Difficulty: 2

ESSAY

23. Consider a motorcycle of mass 150 kg, one wheel of which has a mass of 10 kg and a radius of 30 cm. What is the ratio of the rotational kinetic energy of the wheels to the total translational kinetic energy of the bike? Assume the wheels are uniform disks.

    Answer: 0.133:1 Difficulty: 2

24. A sphere of radius R and mass M is released from rest, and rolls without slipping down a plane inclined at angle G above horizontal. What is the speed of the sphere after it has moved a distance L?

    Answer: 5/6 gL sin G Difficulty: 2

Chapter 8

MULTIPLE CHOICE

25. A hoop of radius 0.5 m and a mass of 0.2 kg is released from rest and allowed to roll down an inclined plane. How fast is it moving after dropping a vertical distance of 3 m?
    a) 2.2 m/s
    b) 3.8 m/s
    c) 5.4 m/s
    d) 7.7 m/s

    Answer: c Difficulty: 2

26. We can best understand how a diver is able to control his rate of rotation while in the air (and thus enter the water in a vertical position) by observing that while in the air,
    a) his total energy is constant.
    b) his kinetic energy is constant.
    c) his potential energy is constant.
    d) his angular momentum is constant.
    e) his linear momentum is constant.

    Answer: d Difficulty: 2

ESSAY

27. A diver can change her rate of rotation in the air by "tucking" her head in and bending her knees. Let's assume that when she is stretched out straight she is rotating at 1 revolution per second. Now she goes into the "tuck and bend", effectively shortening the length of her body by half. What will her rate of rotation be now?

    Answer: 4 Hz Difficulty: 2

MULTIPLE CHOICE

28. A solid cylinder and a hollow cylinder have the same mass and the same radius. Which statement is true concerning their moment of inertia about an axis through the exact center of the flat surfaces?
    a) The hollow cylinder has the greater moment of inertia.
    b) The solid cylinder has the greater moment of inertia.
    c) Both cylinders have the same moment of inertia.
    d) The moment of inertia cannot be determined since it depends on the amount of material removed from the inside of the hollow cylinder.

    Answer: a Difficulty: 2

29. A hollow sphere and a solid sphere of the same mass are spinning with the same angular velocity. Which sphere has the greater angular momentum?
   a) The hollow sphere
   b) The solid sphere
   c) Both spheres have the same angular momentum.
   d) It is impossible to tell without knowing the radii.

   Answer: d Difficulty: 2

30. An ice skater performs a pirouette (a fast spin) by pulling in his outstretched arms close to his body. What happens to his moment of inertia about the axis of rotation?
   a) It does not change.
   b) It increases.
   c) It decreases.
   d) It changes, but it is impossible to tell which way.

   Answer: c Difficulty: 2

31. An ice skater performs a pirouette (a fast spin) by pulling in his outstretched arms close to his body. What happens to his angular momentum about the axis of rotation?
   a) It does not change.
   b) It increases.
   c) It decreases.
   d) It changes, but it is impossible to tell which way.

   Answer: a Difficulty: 2

32. An ice skater performs a pirouette (a fast spin) by pulling in his outstretched arms close to his body. What happens to his rotational kinetic energy about the axis of rotation?
   a) It does not change.
   b) It increases.
   c) It decreases.
   d) It changes, but it is impossible to tell which way.

   Answer: b Difficulty: 2

33. Two uniform solid spheres have the same mass, but one has twice the radius of the other. The ratio of the larger sphere's moment of inertia to that of the smaller sphere is
   a) 4/5
   b) 8/5
   c) 2
   d) 4

   Answer: d Difficulty: 2

Chapter 8

34. Consider two uniform solid spheres where one has twice the mass and twice the diameter of the other. The ratio of the larger moment of inertia to that of the smaller moment of inertia is
a) 2
b) 8
c) 4
d) 10
e) 6

Answer: b Difficulty: 2

35. Consider two uniform solid spheres where both have the same diameter, but one has twice the mass of the other. The ratio of the larger moment of inertia to that of the smaller moment of inertia is
a) 2
b) 8
c) 10
d) 4
e) 6

Answer: a Difficulty: 2

36. An object's angular momentum changes by 10 kg-m^2/s in 2 s. What magnitude average torque acted on this object?
a) 40 N-m
b) 2.5 N-m
c) 20 N-m
d) 5 N-m

Answer: d Difficulty: 2

37. The moment of inertia of a solid cylinder about its axis is given by 0.5MR^2. If this cylinder rolls without slipping, the ratio of its rotational kinetic energy to its translational kinetic energy is
a) 1:1
b) 1:2
c) 2:1
d) 1:3

Answer: b Difficulty: 2

38. Consider two equal mass cylinders rolling with the same translational velocity. The first cylinder (radius = R) is hollow and has a moment of inertia about its rotational axis of MR^2, while the second cylinder (radius = r) is solid and has a moment of inertia about its axis of 0.5 MR^2. What is the ratio of the hollow cylinder's angular momentum to that of the solid cylinder?
a) r^2 to 2R^2
b) 2R^2 to r^2
c) r to 2R
d) 2R to r

Answer: d Difficulty: 2

39. A planet of constant mass orbits the sun in an elliptical orbit. Neglecting any friction effects, what happens to the planet's rotational kinetic energy about the sun's center?
   a) It remains constant.
   b) It increases continually.
   c) It decreases continually.
   d) It increases when the planet approaches the sun, and decreases when it moves farther away.

   Answer: d Difficulty: 2

40. A wheel of moment of inertia of 5.00 kg-m² starts from rest and accelerates under a constant torque of 3.00 N-m for 8.00 s. What is the wheel's rotational kinetic energy at the end of 8.00 s?
   a) 57.6 J
   b) 64.0 J
   c) 78.8 J
   d) 122 J

   Answer: a Difficulty: 2

ESSAY

41. Two children, each of mass 20 kg, ride on the perimeter of a small merry-go-round that is rotating at the rate of 1 revolution every 4 s. The merry-go-round is a disk of mass 30 kg and radius 3 m. The children now both move halfway in toward the center to positions 1.5 m from the axis of rotation. Calculate the kinetic energy before and after, and the final rate of rotation. Explain any changes in energy.

   Answer: K_o = 611 J, K = 1340 J, f = 0.55 Hz. Difficulty: 2

MULTIPLE CHOICE

42. A proton of mass m rotates with an angular speed of 2 x 10⁶ rad/s in a circle of radius 0.8 m in a cyclotron. What is the orbital angular momentum of the proton?
   a) 1.28 x 10⁶ m
   b) 1.76 x 10⁶ m
   c) 3.2 x 10⁶ m
   d) 6.4 x 10⁶ m

   Answer: a Difficulty: 2

Chapter 8

ESSAY

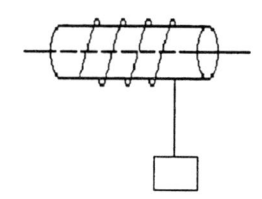

43. A uniform solid cylinder (mass 100 kg and radius 50 cm) is mounted so it is free to rotate about fixed horizontal axis that passes through the centers of its circular ends. A 10-kg block is hung from a massless cord wrapped around the cylinder's circumference. When the block is released, the cord unwinds and the block accelerates downward. What is the block's acceleration?

Answer: 1.63 m/s² Difficulty: 2

MULTIPLE CHOICE

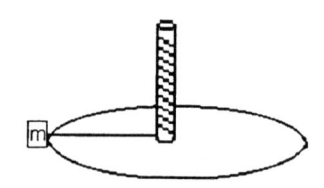

44. Initially, a 2.00-kg mass is whirling at the end of a string (in a circular path of radius 0.750 m) on a horizontal frictionless surface with a tangential speed of 5 m/s. The string has been slowly winding around a vertical rod, and a few seconds later the length of the string has shortened to 0.250 m. What is the instantaneous speed of the mass at the moment the string reaches a length of 0.250 m?
a) 3.9 m/s
b) 15 m/s
c) 75 m/s
d) 225 m/s

Answer: b Difficulty: 2

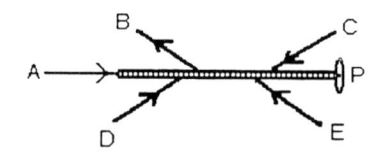

45. Five forces act on a massless rod free to pivot at point P. Which force is producing a counter-clockwise torque about point P?
a) A
b) B
c) C
d) D
e) E

Answer: c Difficulty: 2

46. Which of the following has units of angular momentum?
    a) J/s²
    b) kg-m/s
    c) N-s
    d) kg-m²/s
    e) J-s/kg

    Answer: d Difficulty: 2

47. Consider an object which is subjected to a net torque. That object will experience which of the following?
    a) a constant angular velocity
    b) an angular acceleration
    c) a constant moment of inertia
    d) a linear acceleration and an angular acceleration
    e) an increasing moment of inertia

    Answer: b Difficulty: 1

48. A potter's wheel with amoment of inertia of 8.0 kg-m² has 5.0 N-m applied to it. It starts from rest and 10.s later it has gained what kinetic energy?
    a) 155. J
    b) 118. J
    c) 78. J
    d) 53. J
    e) 44. J

    Answer: c Difficulty: 2

FILL-IN-THE-BLANK

49. A majorette fastens two batons together at their centers to form an X shape. What is the moment of inertia of the combination about an axis through the center of the X if each baton was 1.2 m long and each ball at the end of each baton has a mass of 0.50 kg (assume negligible mass in the rods) ?

    Answer: 0.72 kg-m² Difficulty: 2

50. Consider a bus designed to obtain its motive power from a large rotating flywheel (1400.kg of diameter 1.5m) that is periodically brought up to its maximum speed of 3600.rpm by an electric motor at the terminal. If the bus requires an average power of 12.kilowatts, how long will it operate between recharges?

    Answer: 39. minutes Difficulty: 2

Chapter 8

51. A ballerina spins initially at 1.5 revolutions/second when her arms are extended. She then draws in her arms to her body and her moment of inertia becomes 0.88 kg-m² and her angular speed increases to 4.0 rev/s. Determine her initial moment of inertia.

Answer: 2.3 kg-m² Difficulty: 2

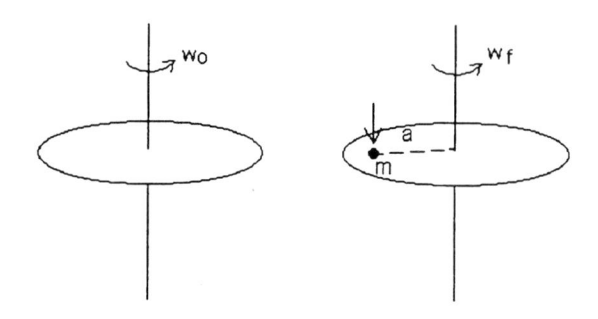

52. A solid disk with diameter 2.00 meters and mass 4.0 kg freely rotates about a vertical axis at 36. rpm. A 0.50 kg hunk of bubblegum is dropped onto the disk and sticks to the disk at a distance d = 80. cm from the axis of rotation.

a) What was the moment of inertia before the gum fell? _____

b) What was the moment of inertia after the gum stuck? _____

c) What is the angular velocity after the gum fell onto the disk ?
_____

Answer: a) 2.0 kg-m²   b) 2.3 kg-m²   c) 31. rpm Difficulty: 2

53. A 50. cm diameter wheel is rotating initially at 2.0 revolutions per second. It slows down uniformly and comes to rest in 15. seconds.

a) What was its angular acceleration? _____

b) Through how many revolutions did it turn in those 15. s? _____

c) What was the centripetal acceleration at the instant when it was rotating initially? _____

d) What was the tangential acceleration? _____

Answer: a) -0.13 rev/s² = -0.84 rad/s²   b) 15. rev   c) 39. m/s²   d) -0.21 m/s²

Difficulty: 2

54. A girl weighing 450.N sits on one end of a seesaw that is 3.0 m long and is pivoted 1.3 m from the child. If the seesaw is just balanced when a boy sits at the opposite end, what is his weight? Neglect the weight of the seesaw.

Answer: 344. Newtons   Difficulty: 2

TRUE/FALSE

55. A pencil balanced on its tip is in stable equilibrium as long as it does not move (stays balanced).

Answer: F   Difficulty: 1

FILL-IN-THE-BLANK

56. Suppose a star like our sun were rotating once a month, and it collapsed to form a "white dwarf" (about a thousand times smaller diameter). Estimate its rotation velocity.

Answer: approx 23.rpm = 2.4 radians/s   Difficulty: 2

MULTIPLE CHOICE

57. A ball, solid cylinder and a hollow pipe all have equal masses and radii. If the three are released simultaneously at the top of an inclined plane, which will reach the bottom first?
a) ball
b) cylinder
c) pipe
d) none of the above

Answer: a   Difficulty: 2

TRUE/FALSE

58. If there is a net force on a system, the angular momentum cannot be conserved.

Answer: F   Difficulty: 2

FILL-IN-THE-BLANK

59. Consider a car with its wheels rolling (not sliding) on the road. How fast must the auto move for some point on a wheel to reach the speed of sound? Explain.

Answer:
Half the speed of sound. The bottom of the wheel has not linear speed, the middle has the car's speed, and the top of the wheel is moving at twice the axis speed.
Difficulty: 2

Chapter 8

MULTIPLE CHOICE

60. A planet speeds up in its orbit as it gets closer to the sun because of:
a) the torque exerted on the planet by the Sun.
b) conservation of energy.
c) conservation of momentum.
d) conservation of angular momentum.

Answer: d Difficulty: 2

FILL-IN-THE-BLANK

61. Is it easier to swing a bat holding the handle at the end or "choked up"? Why?

Answer:
"Choked up" because the moment of inertia is less about the axis of rotation (bat center of mass moved toward axis).

Difficulty: 1

MULTIPLE CHOICE

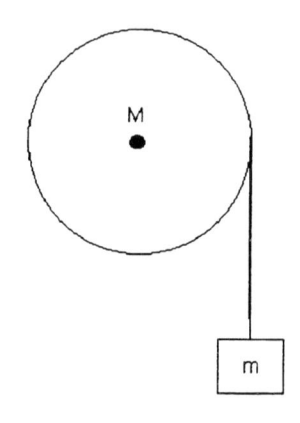

62. A 4.0 kg mass is hung from a string which is wrapped around a cylindrical pulley (see figure). If the mass accelerates downward at 4.90 m/s^2, what is the mass of the pulley?
a) 2.0kg
b) 4.0kg
c) 6.0kg
d) 8.0kg
e) 10.kg

Answer: d Difficulty: 2

FILL-IN-THE-BLANK

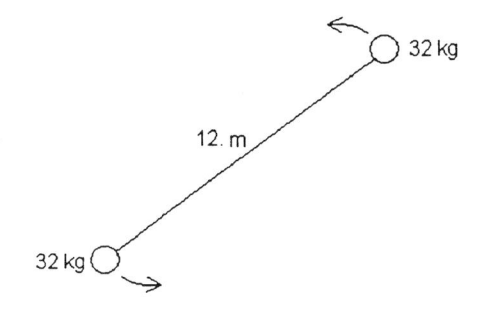

63. NASA puts a satellite in space which consists of two (32.kg each) masses connected by a 12. meter long light cable. The entire system initially is rotating 4.0 rpm about an axis (at the center of mass) perpendicular to the cable. Motors in each mass reel out more cable so that the masses double their separation (to 24.m).
    a) What is the final ROTATIONAL VELOCITY?
    b) The final KINETIC ENERGY is what fraction of the initial?

    Answer: a) 1.0 rpm   b) 1/4   Difficulty: 2

64. Peter swings a ball, on the end of a 110.cm string, in a vertical circle (radius 110.cm).
    a) what minimum speed is needed by the ball to keep it moving on the arc of the circle when it is at the top of the circular loop?
    b) at that minimum speed, what is the tension in the string at the top?

    Answer: a) 3.28 m/s   b) zero Difficulty: 2

65. A comet is observed to orbit a star on an ellipse. When closest to the star (0.25 AU) it is moving 48.km/s. How fast is it moving when it reaches its farthest point from the star (60.AU distance)?

    Answer: 0.20 km/s Difficulty: 2

Chapter 8

ESSAY

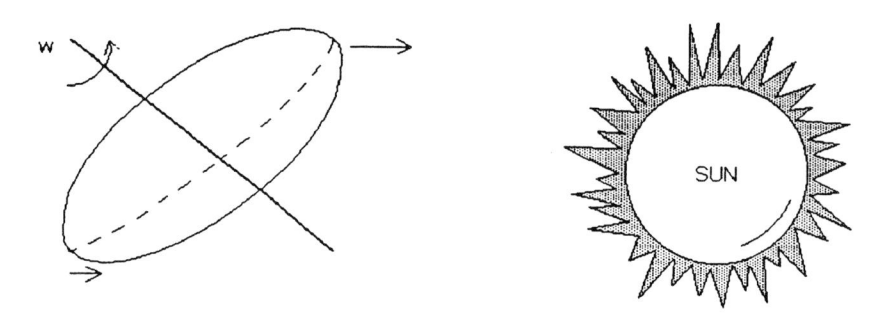

66. Like a gyroscope, the Earth precesses with a 26,000-year period. Explain how the Sun and Moon can exert torques on the Earth to cause this.

Answer:
The Earth bulges at the equator and is tilted so the differential gravitational attraction on the bulges exerts a torque (see figure).

Difficulty: 2

Chapter 9

MATCHING

Match the unit to the physical quantity.

a) m³/s
b) dimensionless
c) m²/N
d) N/m²
e) N/m
f) torr
g) 14.7 lb/ft²
h) Pa-s
i) poise

1. flow rate

    Answer: a) m³/s Difficulty: 2

2. Reynolds number

    Answer: b) dimensionless Difficulty: 2

3. compressibility

    Answer: c) m²/N Difficulty: 2

4. Young's modulus

    Answer: d) N/m² Difficulty: 2

5. surface tension

    Answer: e) N/m Difficulty: 2

6. 1 mm Hg

    Answer: f) torr Difficulty: 2

7. 1 atmosphere

    Answer: g) 14.7 lb/ft² Difficulty: 2

8. SI unit for viscosity

    Answer: h) Pa-s Difficulty: 2

9. cgs unit for viscosity

    Answer: i) poise Difficulty: 2

Chapter 9

MULTIPLE CHOICE

10. All of the following are dimensionless except
    a) Reynolds number.
    b) specific gravity.
    c) strain.
    d) bulk modulus.

    Answer: d Difficulty: 2

11. Stress is
    a) the strain per unit length.
    b) the same as force.
    c) the ratio of the change in length.
    d) applied force per cross-sectional area.

    Answer: d Difficulty: 2

12. Strain is
    a) the ratio of the change in length to the original length.
    b) the stress per unit area.
    c) the applied force per unit area.
    d) the ratio of stress to elastic modulus.

    Answer: a Difficulty: 2

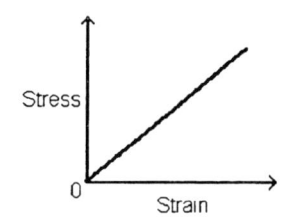

13. The slope of the straight line shown on the graph is called the
    a) bulk modulus.
    b) Young's modulus.
    c) compressibility.
    d) pressure.

    Answer: b Difficulty: 2

ESSAY

14. A mass of 50 kg is suspended from a steel wire of diameter 1 mm and length 11.2 m. How much will the wire stretch? The Young's modulus for steel is 20×10^{10} N/m².

    Answer: 3.7 mm Difficulty: 2

15. At a depth of about 1030 m in the sea the pressure has increased by 100 atmospheres (to about 10^7 N/m^2). By how much has 1 m^3 of water been compressed by this pressure? The bulk modulus of water is 2.3 x 10^9 N/m^2

   Answer: 0.0043 m^3 Difficulty: 2

MULTIPLE CHOICE

16. The weight that will cause a wire of diameter d to stretch a given distance, for a fixed length of wire, is
    a) independent of what the length of the wire is.
    b) proportional to d.
    c) proportional to d^2.
    d) independent of d.

    Answer: c   Difficulty: 2

ESSAY

17. What is the pressure of a submarine at a depth of 100 m?

    Answer: 9.8 x 10^5 Pa   Difficulty: 2

18. A window for viewing fish in the local aquarium is square and measures 0.8 m x 0.8 m. Its center is 12 m below the surface. What is the total force on the window?

    Answer: 75,300 N   Difficulty: 2

19. Suppose that in an examination of a spinal injury it is found that the spinal fluid (with about the same density as water) will rise a vertical height of 13 cm in an open tube. What pressure does this correspond to, expressed in mm of mercury and in atmospheres?

    Answer: 9.56 mm Hg or 0.013 atm   Difficulty: 2

MULTIPLE CHOICE

20. When you are scuba diving, the pressure on your face plate
    a) will be greatest when you are facing upward.
    b) will be greatest when you are facing downward.
    c) depends only on your depth, and not on how you are oriented.
    d) is independent of both depth and orientation.

    Answer: d   Difficulty: 2

Chapter 9

21. A plastic block of dimensions 2 cm x 3 cm x 4 cm has a mass of 30 g. What is its density?
    a) 0.8 g/cm^3
    b) 1.2 g/cm^3
    c) 1.25 g/cm^3
    d) 1.60 g/cm^3

    Answer: c Difficulty: 2

22. Suppose that an 80-kg person walking on crutches supports all his weight on the two crutch tips, each of which is circular with a diameter of 1". What pressure is exerted on the floor? Express the answer in psi.
    a) 56 psi
    b) 112 psi
    c) 142 psi
    d) 224 psi

    Answer: d Difficulty: 2

23. A circular window of 30 cm diameter in a submarine can withstand a maximum force of 1.88 x 10^6 N. What is the maximum depth to which the submarine can go without damaging the window?
    a) 680 m
    b) 750 m
    c) 1200 m
    d) 1327 m

    Answer: a Difficulty: 2

24. A copper wire is found to break when subjected to minimum tension of 36 N. If the wire diameter were half as great, we would expect the wire to break when subjected to a minimum tension of
    a) 9 N
    b) 18 N
    c) 25.5 N
    d) 36 N
    e) 50.7 N

    Answer: e Difficulty: 2

25. Substance A has a density of 3 g/cm^3 and substance B has a density of 4 g/cm^3. In order to obtain equal masses of these two substances, the ratio of the volume of A to the volume of B will be equal to
    a) 1:3
    b) 4:3
    c) 3:4
    d) 1:4

    Answer: b Difficulty: 2

26. A piece of iron sinks to the bottom of a lake where the pressure is 21 atm. Which statement best describes what happens to the volume of that piece of iron?
a) Its volume decreases slightly.
b) Its volume becomes 21 times greater.
c) Its volume increases slightly.
d) There has been no change in the volume of the iron.

Answer: c Difficulty: 2

27. A piece of iron sinks to the bottom of a lake where the pressure is 21 atm. Which statement best describes what happens to the density of that piece of iron?
a) Its density decreases slightly.
b) Its density becomes 21 times greater.
c) Its density increases slightly.
d) There has been no change in the density of the iron.

Answer: c Difficulty: 2

ESSAY

28. A cylindrical rod of length 12 cm and diameter 2 cm will just barely float in water. What is its mass?

Answer: 37.7 g Difficulty: 2

29. A crane lifts a steel submarine (density = 7.8 x 10^3 kg/m^3) of mass 20,000 kg. What is the tension in the lifting cable (a) when the submarine is submerged, and (b) when it is entirely out of the water?

Answer: (a) 1.71 x 10^5 N; (b) 1.96 x 10^5 N Difficulty: 2

TRUE/FALSE

30. An object completely submerged in water must either rise or fall.

Answer: F Difficulty: 2

ESSAY

31. A polar bear of mass 200 kg stands on an ice floe 100 cm thick. What is the minimum area of the floe that will just support the bear in saltwater of specific gravity 1.03? The specific gravity of ice is 0.98.

Answer: 4000 m^3 Difficulty: 2

Chapter 9

32. Oil of specific gravity 0.90 is poured on top of a beaker of water. A cube of plastic of specific gravity 0.95 is placed in the liquid. What fraction of the cube will be immersed in water?

Answer: 0.5 Difficulty: 2

MULTIPLE CHOICE

33. A sunken steel ship has a mass of 500,000 kg. It is filled with water. In order to lift the ship, air bags are to be inflated inside the hull. What volume of air is needed? The specific gravity of steel is 7.8.
   a) 225 m^3
   b) 436 m^3
   c) 1266 m^3
   d) 2778 m^3

Answer: b Difficulty: 2

34. A container of water is placed on a scale, and the scale reads 120 g. Now a 20-g piece of copper (specific gravity = 8.9) is suspended from a thread and lowered into the water, not touching the bottom of the container. What will the scale now read?
   a) 120 g
   b) 122 g
   c) 138 g
   d) 140 g

Answer: b Difficulty: 2

TRUE/FALSE

35. When a glass tube is inserted into mercury (Hg), the mercury level inside the tube is lower than the mercury level outside the tube.

Answer: T Difficulty: 2

ESSAY

36. The surface tension of water is 0.073 N/m. How high will water rise in a capillary tube of diameter 1.2 mm?

Answer: 2.45 cm Difficulty: 2

MULTIPLE CHOICE

37. When a tube of diameter d is placed in water, the water rises to a height h. If the diameter were half as great, how high would the water rise?
   a) h/2
   b) h
   c) 2h
   d) 4h

   Answer: c Difficulty: 2

TRUE/FALSE

38. If a liquid does not wet the inner surface of a capillary tube inserted into it, the cohesion of the liquid is greater than the adhesion between the liquid and surface.

   Answer: T Difficulty: 2

MULTIPLE CHOICE

39. Consider a rectangular frame of length 0.120 m and width 0.0600 m with a soap film formed within its confined area. If the surface tension of soapy water is 0.0260 N/m, how much force does the soap film exert on the 0.600 m side?
   a) 0.00624 N
   b) 0.00156 N
   c) 0.00312 N
   d) None of the above.

   Answer: c Difficulty: 2

40. A narrow tube is placed vertically in a pan of water, and the water rises in the tube to 4 cm above the level of the pan. The surface tension in the liquid is lowered to one-half its original value by the addition of some soap. What happens to the height of the liquid column in the tube?
   a) It drops to 2 cm.
   b) It drops to 1 cm.
   c) It drops or rises, but the height cannot be calculated from the information given.
   d) It remains at the same height.

   Answer: a Difficulty: 2

TRUE/FALSE

41. Falling drops of milk tend to form spheres, in order to maximize their surface area.

   Answer: T Difficulty: 2

Chapter 9

MULTIPLE CHOICE

42. When soup gets cold, it often tastes greasy. This "greasy" taste seems to be associated with oil spreading out all over the surface of the soup, instead of staying in little globules. To us "physikers", this is readily explained in terms of
    a) the Bernoulli effect.
    b) Archimedes Principle.
    c) the decrease in the surface tension of water with increasing temperature.
    d) the increase in the surface tension of water with increasing temperature.

    Answer: d Difficulty: 2

ESSAY

43. A hole of radius 1 mm occurs in the bottom of a water storage tank that holds water at a depth of 15 m. At what rate will water flow out of the hole?

    Answer: 0.053 L/s Difficulty: 2

MULTIPLE CHOICE

44. Water flows through a horizontal pipe of cross-sectional area 10 cm^2 at a pressure of 0.25 atm. The flow rate is 1 L/s. At a valve, the effective cross-sectional area of the pipe is reduced to 5 cm^2. What is the pressure at the valve?
    a) 0.112 atm
    b) 0.157 atm
    c) 0.200 atm
    d) 0.235 atm

    Answer: d Difficulty: 2

ESSAY

45. Suppose that the build-up of fatty tissue on the wall of an artery decreased the radius by 10%. By how much would the pressure provided by the heart have to be increased to maintain a constant blood flow?

    Answer: 52% Difficulty: 2

MULTIPLE CHOICE

46. SAE No. 10 oil has a viscosity of 0.2 Pa-s. How long would it take to pour 4 L of oil through a funnel with a neck 15 cm long and 2 cm in diameter? Assume the surface of the oil is kept 6 cm above the top of the neck, and neglect any drag effects due to the upper part of the funnel.
    a) 46 s
    b) 52 s
    c) 84 s
    d) 105 s

    Answer: b Difficulty: 2

TRUE/FALSE

47. Air is forced to flow over an object whose cross-section is shown. The speed of the air flowing close to the top surface is greater than the speed of the air flowing close to the bottom surface.

    Answer: F Difficulty: 2

MULTIPLE CHOICE

48. When a small spherical rock of radius r falls through water, it experiences a drag force (a)(r)(v), where "v" is its velocity and "a" is a constant proportional to the viscosity of water. From this, one can deduce that if a rock of diameter 2 mm falls with terminal velocity, "v", then a rock of diameter 4 mm will fall with terminal velocity
    a) v
    b) v/2
    c) 2 v
    d) 4 v

    Answer: d Difficulty: 2

49. Two Styrofoam balls, of radii R and 2R, are released simultaneously from a tall tower. Which will reach the ground first?
    a) Both will reach the ground simultaneously.
    b) The larger one
    c) The smaller one
    d) The result will depend on the atmospheric pressure.

    Answer: b Difficulty: 2

Chapter 9

TRUE/FALSE

50. When a baseball pitcher throws a curve ball, he makes the ball spin about an axis perpendicular to the direction in which the ball's center of mass is moving.

    Answer: T Difficulty: 2

MULTIPLE CHOICE

51. Which has the greatest effect on the flow of fluid through a pipe? That is, if you made a 10% change in each of the quantities below, which would cause the greatest change in the flow rate?
    a) The fluid viscosity
    b) The pressure difference
    c) The length of the pipe
    d) The radius of the pipe

    Answer: d Difficulty: 2

FILL-IN-THE-BLANK

52. Consider a submarine at a depth of 67.meters in sea water (density 1.03 grams/cc) and the interior of the sub is at 1.0 atmosphere.

    a) What is the outside pressure at that depth?

    b) find the total force that must be withstood by a square hatch (0.90 m square) which leads to the sub interior.

    Answer: a) 6.76×10^5 Pa = 6.68 atm.   b) 466.kN   Difficulty: 2

53. An iron barge 25.m long and 4.0 m wide has a mass of 100,000.kg. What minimum depth of water is needed to float it?

    Answer: 1.0 meter Difficulty: 2

MULTIPLE CHOICE

54. A submarine rests on the sea bottom. The normal force exerted up on the sub by the sea-floor is equal to:
    a) the weight of the sub
    b) the weight of the displaced water
    c) the weight of the sub plus weight of displaced water
    d) the weight of the sub less the weight of the displaced water
    e) the buoyant force less 1 atmosphere acting on the sub

    Answer: d Difficulty: 2

ESSAY

55. Consider ice cubes floating in a glass of water. What will happen to the level of the water as the ice melts?

Answer:
The water level remains unchanged since the ice displaces its own weight of water

Difficulty: 2

MULTIPLE CHOICE

56. Atmospheric pressure does **NOT** correspond to approximately:
    a) 2.4×10^2 oz/in^2
    b) 1.0×10^2 N/m^2
    c) 2.2×10^3 lb/ft^2
    d) 0.1 MPa

Answer: b Difficulty: 1

57. Which one of the following would be expected to have the smallest BULK MODULUS?
    a) Helium vapor
    b) solid Iron
    c) solid Uranium
    d) liquid Water

Answer: a Difficulty: 1

FILL-IN-THE-BLANK

58. What is the maximum depth from which a pump on the surface can "suck" up water?

Answer: Atmospheric pressure will support a column of water about 10.m high.

Difficulty: 2

MULTIPLE CHOICE

59. The Reynolds number does NOT depend upon:
    a) density
    b) viscosity
    c) flow speed
    d) pressure

Answer: d Difficulty: 1

Chapter 9

60. The pressure differential across a wing cross section due to the difference in air flow is explained by:
   a) Archimedes Principle
   b) Bernoulli's equation
   c) Newton's third law
   d) Poiseuille's law

   Answer: b Difficulty: 1

61. "Pressure applied to an enclosed fluid is transmitted to every point in the fluid and to the enclosure walls" is known as:
   a) Fermet's principle
   b) Pascal's principle
   c) Archimedes' principle
   d) Bernoulli's principle

   Answer: b Difficulty: 2

62. Which of the following is NOT a unit of pressure?
   a) psi
   b) inches of Mercury
   c) atmosphere
   d) Pascal
   e) $N-m^2$

   Answer: e Difficulty: 1

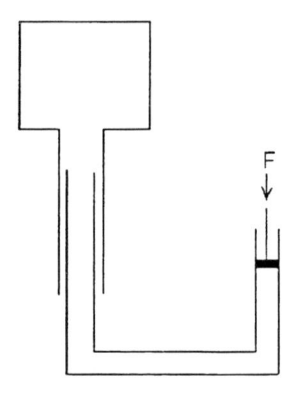

63. Instead of cables, a hydraulic lift raises an elevator weighing 2.5 kilonewtons. The input piston has a 2.0cm diameter and the lift piston has a 28.cm diameter. What minimum force must be applied to the input piston?

   a) 13.N
   b) 179.N
   c) 2.5 kN
   d) 35.kN
   e) 490.kN

   Answer: a Difficulty: 2

150

64. A mercury barometer
   a) is a closed tube manometer.
   b) is an open tube manometer.
   c) measures gauge pressure.
   d) is usually 760.cm tall.

   Answer: a   Difficulty: 1

ESSAY

65. When a heavy barge floats above a fish swimming in the water below, does the PRESSURE change exerted upon the fish (and why)?

   Answer:
   NO. The barge displaces just enough water to equal its weight. The barge can be replaced by water up to the water surface and the weight of that overlying material is the same upon the fish.
   Difficulty: 2

66. An object is submerged in a swimming pool and of course the water exerts a buoyant force on the object. Suppose a thousand gallons of oil is spilled into the pool and now floats above the water. How is the buoyant force on the object changed by the overlying oil?

   Answer:
   No change at all as long as the density of the water remains unchanged. After all, the buoyant force depends upon the volume displaced and the DENSITY of the water.
   Difficulty: 2

MULTIPLE CHOICE

67. A spherical inflated balloon is submerged in a pool of water. If it is further inflated so that is radius doubles, how is the buoyant force affected?
   a) not at all
   b) 2 times larger
   c) 4 times larger
   d) 6 times larger
   e) 8 times larger
   f) need to know how weight of balloon changed.

   Answer: e   Difficulty: 2

Chapter 9

FILL-IN-THE-BLANK

68. A spherically shaped hot air balloon of diameter 20.meters, just floats (hovers) when the hot air inside has been heated to a density of 1.00 kg/m^3. What is the weight of the balloon and cargo (not including the air inside) if the suroounding air has a density of 1.30 kg/m^3?

    Answer: 1.2 x 10^4 Newtons (weight of 1.3x10^3 kg) Difficulty: 2

69. A long telephone pole (wood density 0.80 kg/m^3) is vertically lowered into a slightly larger vertical pipe which is filled with salt water of density 1.04 kg/m^3. The pole ends up floating with what percent of the pole above the water?

    Answer: 23.% above (77% submerged) Difficulty: 2

MULTIPLE CHOICE

70. COHESIVE FORCES are
    a) repulsive forces between unlike molecules
    b) attractive forces between unlike molecules
    c) repulsive forces between like molecules
    d) attractive forces between like molecules

    Answer: d Difficulty: 2

FILL-IN-THE-BLANK

71. An incompressible fluid flows at 0.252 m/s through a 44.mm diameter (circular cross section) pipe. The pipe widens to a square cross sectional area 5.5 cm on a side. Assuming steady flow:
    a) What is the speed through the "square" section?
    b) What is the flow rate in liters/minute?

    Answer: a) 0.13 m/s   b) 23. liters/minute Difficulty: 2

MULTIPLE CHOICE

72. The density of water is
    a) 1.0 kg/m^3
    b) 1.0 Mg/m^3
    c) 1.0 x 10^3 kg/cm^3
    d) 1.0 x 10^1 kg/cm^3
    e) 1.0 x 10^{-3} g/m^3

    Answer: b Difficulty: 1

Chapter 10

MATCHING

Match the unit to the physical quantity.

a) 273.16° K   d) 373.16° K   g) J/K
b) 1/°C   e) molecules/mole   h) joules
c) J/mol°K   f) 1 C°

1. temperature of triple point of water

   Answer: a) 273.16° K Difficulty: 2

2. coefficient of volume expansion

   Answer: b) 1/°C Difficulty: 2

3. universal gas constant

   Answer: c) J/mol°K Difficulty: 2

4. boiling point of water

   Answer: d) 373.16° K Difficulty: 2

5. Avogadro's number

   Answer: e) molecules/mole Difficulty: 2

6. 9/5 of a Fahrenheit degree

   Answer: f) 1 C° Difficulty: 2

7. Boltzmann's constant

   Answer: g) J/K Difficulty: 2

8. internal energy

   Answer: h) joules Difficulty: 2

MULTIPLE CHOICE

9. Which temperature scale never gives negative temperatures?
   a) Kelvin
   b) Fahrenheit
   c) Celsius

   Answer: a Difficulty: 2

Chapter 10

10. Which two temperature changes are equivalent?
    a) 1 K° = 1 F°
    b) 1 F° = 1 C°
    c) 1 C° = 1 K°

    Answer: c Difficulty: 2

11. The temperature in your classroom is approximately
    a) 68° K
    b) 68° C
    c) 50° C
    d) 295° K

    Answer: d Difficulty: 2

12. Express 45° C in ° F.
    a) 25° F
    b) 57° F
    c) 81° F
    d) 113° F

    Answer: d Difficulty: 2

ESSAY

13. At what temperature are the numerical readings on the Fahrenheit and Celsius scales the same?

    Answer: -40° Difficulty: 2

MULTIPLE CHOICE

14. Express your body temperature (98.6° F) in Celsius degrees.
    a) 37° C
    b) 45.5° C
    c) 66.6° C
    d) 72.6° C

    Answer: a Difficulty: 2

15. A temperature change of 20° C corresponds to a temperature change of
    a) 68° F
    b) 11.1° F
    c) 36° F
    d) none of the above

    Answer: c Difficulty: 2

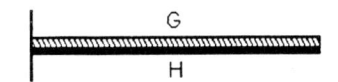

16. A bimetallic strip, consisting of metal G on the top and metal H on the bottom, is rigidly attached to a wall at the left. The coefficient of linear thermal expansion for metal G is greater than that of metal H. If the strip is uniformly heated, it will
    a) curve upward.
    b) curve downward.
    c) remain horizontal, but get longer.
    d) bend in the middle.

    Answer: b Difficulty: 2

ESSAY

17. A container of an ideal gas at 1 atm is compressed to one-third its volume, with the temperature held constant. What is its final pressure?

    Answer: 3 atm Difficulty: 2

MULTIPLE CHOICE

18. In order to double the average speed of the molecules in a given sample of gas, the temperature (measured in Kelvins) must
    a) quadruple.
    b) reduce to one-fourth its original value.
    c) reduce to one-half its original value.
    d) triple.
    e) double.

    Answer: a Difficulty: 2

19. How many water molecules are there in 36 g of water? Express your answer as a multiple of Avogadro's number N_A. (The molecular structure of a water molecule is H_2O.)
    a) 36 N^A
    b) 2 N^A
    c) 18 N^A
    d) None of the above

    Answer: b Difficulty: 2

ESSAY

20. How many mol are there in 2 kg of copper? (The atomic weight of copper is 63.5 and its specific gravity is 8.9.)

    Answer: 31.5 mol Difficulty: 2

Chapter 10

21. A constant pressure gas thermometer is initially at 28°C. If the volume of gas increases by 10%, what is the final Celsius temperature?

Answer: 630°C Difficulty: 2

MULTIPLE CHOICE

22. The temperature of an ideal gas increases from 2°C to 4°C while remaining at constant pressure. What happens to the volume of the gas?
a) It decreases slightly.
b) It decreases to one-half its original volume.
c) It more than doubles.
d) It doubles.
e) It increases slightly.

Answer: e Difficulty: 2

23. Both the pressure and volume of a given sample of an ideal gas double. This means that its temperature in Kelvins must
a) double.
b) quadruple.
c) reduce to one-fourth its original value.
d) remain unchanged.
e) reduce to one-half its original value.

Answer: b Difficulty: 2

24. If the pressure acting on an ideal gas at constant temperature is tripled, its volume is
a) reduced to one-third.
b) increased by a factor of three.
c) increased by a factor of two.
d) reduced to one-half.
e) none of the other choices is correct.

Answer: a Difficulty: 2

ESSAY

25. Two liters of a perfect gas are at 0°C and 1 atm. If the gas is nitrogen, N_2, determine (a) the number of mol (b) the number of molecules (c) the mass of the gas.

Answer: (a) 0.089 mol, (b) 5.34×10^{22}, (c) 2.5 g Difficulty: 2

26. What is the average separation between air molecules at STP?

Answer: 3.34×10^{-7} cm Difficulty: 2

MULTIPLE CHOICE

27. A sample of a diatomic ideal gas occupies 33.6 L under standard conditions. How many mol of gas are in the sample?
    a) 0.75
    b) 3.0
    c) 1.5
    d) None of the above

    Answer: c Difficulty: 2

28. A container is filled with a mixture of helium and oxygen gases. A thermometer in the container indicates that the temperature is 22°C. Which gas molecules have the greater average speed?
    a) The helium molecules do because they are monatomic.
    b) It is the same for both because the temperatures are the same.
    c) The oxygen molecules do because they are more massive.
    d) The helium molecules do because they are less massive.
    e) The oxygen molecules do because they are diatomic.

    Answer: d Difficulty: 2

29. A container is filled with a mixture of helium and oxygen gases. A thermometer in the container indicates that the temperature is 22°C. Which gas molecules have the greater average kinetic energy?
    a) It is the same for both because the temperatures are the same.
    b) The oxygen molecules do because they are diatomic.
    c) The helium molecules do because they are less massive.
    d) The helium molecules do because they are monatomic.
    e) The oxygen molecules do because they are more massive.

    Answer: a Difficulty: 2

30. Oxygen molecules are 16 times more massive than hydrogen molecules. At a given temperature, the average molecular kinetic energy of oxygen, compared to hydrogen
    a) is greater.
    b) is less.
    c) is the same.
    d) cannot be determined since pressure and volume are not given.

    Answer: c Difficulty: 2

31. Oxygen molecules are 16 times more massive than hydrogen molecules. At a given temperature, how do their average molecular speeds compare? The oxygen molecules are moving
    a) 4 times faster.
    b) at 1/4 the speed.
    c) 16 times faster.
    d) at 1/16 the speed.

    Answer: b Difficulty: 2

Chapter 10

32. A sample of an ideal gas is slowly compressed to one-half its original volume with no change in temperature. What happens to the average speed of the molecules in the sample?
   a) It does not change.
   b) It doubles.
   c) It halves.
   d) None of the above.

   Answer: a Difficulty: 2

33. A sample of an ideal gas is heated and its Kelvin temperature doubles. What happens to the average speed of the molecules in the sample?
   a) It does not change.
   b) It doubles.
   c) It halves.
   d) None of the above.

   Answer: d Difficulty: 2

34. The number of molecules in one mole a substance
   a) depends on the molecular weight of the substance.
   b) depends on the atomic weight of the substance.
   c) depends on the density of the substance.
   d) is the same for all substances.

   Answer: d Difficulty: 2

ESSAY

35. A steel cable is 20 m long when the temperature is 20° C. What will be its length when the temperature drops to 0° C? (The coefficient of thermal expansion of steel is 12×10^{-6} $(°K)^{-1}$).

   Answer: 19.9952 m Difficulty: 2

MULTIPLE CHOICE

36. The coefficient of linear expansion for aluminum is 1.80×10^{-6} $(°K)^{-1}$. What is its coefficient of volume expansion?
   a) 6×10^{-6} $(°K)^{-1}$.
   b) 8×10^{-18} $(°K)^{-1}$.
   c) 5×10^{-6} $(°K)^{-1}$.
   d) 8×10^{-6} $(°K)^{-1}$.
   e) None of the above choices is correct.

   Answer: a Difficulty: 2

ESSAY

37. A mercury thermometer has a bulb of volume 0.100 cm³ at 10°C. The capillary tube above the bulb has a cross-sectional area of 0.012 mm². The volume thermal expansion coefficient of mercury is 1.80 x 10⁻⁴ (°K)⁻
    How much will the mercury rise when the temperature rises by 30°C?

    Answer: 3 mm   Difficulty: 2

MULTIPLE CHOICE

38. When the engine of your car heats up, the spark plug gap will
    a) increase.
    b) decrease.
    c) remain unchanged.
    d) decrease at first and then increase later, so that the two effects cancel once the engine reaches operating temperature.
    e) none of the above is true.

    Answer: a   Difficulty: 2

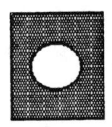

39. Consider a flat steel plate with a hole through its center. When the plate's temperature is increased, the hole will
    a) expand only if it takes up more than half the plate's surface area.
    b) contract if it takes up less than half the plate's surface area.
    c) always contract.
    d) always expand.

    Answer: d   Difficulty: 2

40. By how much will a slab of concrete 18 m long contract when the temperature drops from 24°C to -16°C? (The coefficient of linear thermal expansion for concrete is 10⁻⁵ per degree C.)
    a) 0.5 cm
    b) 0.7 cm
    c) 1.2 cm
    d) 1.5 cm

    Answer: b   Difficulty: 2

Chapter 10

41. A bolt hole in a brass plate has a diameter of 1.2 cm at 20°C. What is the diameter of the hole when the plate is heated to 220°C? (The coefficient of linear thermal expansion for brass is 19 x 10^{-6} per degree C.)
   a) 1.205 cm
   b) 1.195 cm
   c) 1.200 cm
   d) 1.210 cm

   Answer: a   Difficulty: 2

TRUE/FALSE

42. When a lake freezes during the winter, the water at the bottom of the lake is at a higher temperature than the water above it.

   Answer: T   Difficulty: 2

MULTIPLE CHOICE

43. The surface water temperature on a large, deep lake is 3°C. A sensitive temperature probe is lowered several m into the lake. What temperature will the probe record?
   a) A temperature warmer than 3°C
   b) A temperature less than 3°C
   c) A temperature equal to 3°C

   Answer: a   Difficulty: 2

44. The volume coefficient of thermal expansion for gasoline is 950 x 10^{-6} per degree C. By how much does the volume of 1 L of gasoline change when the temperature rises from 20°C to 40°C?
   a) 6 cm^3
   b) 12 cm^3
   c) 19 cm^3
   d) 37 cm^3

   Answer: c   Difficulty: 2

ESSAY

45. Heat is added to an ideal gas at 20°C. If the internal energy of the gas increases by a factor of three during this process, what is the final temperature?

   Answer: 606°C   Difficulty: 2

46. At what temperature is the average kinetic energy of an atom in helium gas equal to 6.21 x 10^{-21} J?

   Answer: 300°K   Difficulty: 2

MULTIPLE CHOICE

47. Ten joules of heat energy are transferred to a sample of ideal gas at constant pressure. As a result, the internal energy of the gas
    a) increases by 10 J.
    b) increases by less than 10 J.
    c) increases by more than 10 J.
    d) remains unchanged.

    Answer: b Difficulty: 2

48. Ten joules of heat energy are transferred to a sample of ideal gas at constant volume. As a result, the internal energy of the gas
    a) increases by 10 J.
    b) increases by less than 10 J.
    c) increases by more than 10 J.
    d) remains unchanged.

    Answer: a Difficulty: 2

49. An ideal gas with internal energy U at 200°C is heated to 400°C. Its internal energy then will be
    a) still U.
    b) 2 U.
    c) 1.4 U.
    d) 1.2 U.

    Answer: c Difficulty: 2

50. An ideal gas at STP is first compressed until its volume is half the initial volume, and then it is allowed to expand until its pressure is half the initial pressure. All of this is done while holding the temperature constant. If the initial internal energy of the gas is U, the final internal energy of the gas will be
    a) U.
    b) U/3.
    c) U/2.
    d) 2U.

    Answer: a Difficulty: 2

51. The internal energy of an ideal gas depends on
    a) its volume.
    b) its pressure.
    c) its temperature.
    d) all of the above.

    Answer: c Difficulty: 2

Chapter 10

FILL-IN-THE-BLANK

52. Ten meter long steel railroad rails are laid end to end, with no space between, on a hot day when the temperature was 115.°F. Six months later the temperature has dropped to -2.°F and the rails now have how much space between each?
    [linear expansion coef. for steel is $12.\times10^{-6}$ C^{-1}]

    Answer: 0.78 cm Difficulty: 2

53. The temperature is such that the rms speed of Nitrogen molecules is 900.m/s. at 750.mm Hg pressure. If the pressure increases to 780.mm Hg while the temperature remains constant, what is the rms speed now?
    [molecular weight of Nitrogen molecule is 28.]

    Answer: same 900.m/s since depends only on temperature Difficulty: 2

MULTIPLE CHOICE

54. Average body temperature is usually said to be:
    a) 31.°C
    b) 33.°C
    c) 35.°C
    d) 37.°C
    e) 39.°C

    Answer: d Difficulty: 1

55. Aluminum has a positive coefficient of thermal expansion. Consider a round hole that has been drilled in a large sheet of Aluminum. As the temperature increases and the surrounding metal expands, the hole diameter will:
    a) increase
    b) decrease
    c) remain constant
    d) depends how much metal surrounds the hole

    Answer: a Difficulty: 2

56. Which of the following would correctly convert a Fahrenheit temperature (F) to a Kelvin (K) value?
    a) K = (5/9)F + 255
    b) K = (5/9)(F - 32)
    c) K = (9/5)(F - 32) + 273
    d) K = (9/5) F + 273
    e) K = (5/9)F - 273

    Answer: a Difficulty: 2

57. On a winter -3.°C morning, the pressure in a car tire reads 28.psi on a pressure gauge. After several hours of hiway traveling, the pressure reads 35.psi. Assuming the volume remains constant and normal atmospheric pressure of 15.psi, what is the tire temperature?
   a) 15.°C
   b) 25.°C
   c) 35.°C
   d) 45.°C
   e) 55.°C

   Answer: c Difficulty: 2

FILL-IN-THE-BLANK

58. "Absolute Zero" is how many Fahrenheit degrees below freezing (water)?

    Answer: 491 F° below "freezing" Difficulty: 1

MULTIPLE CHOICE

59. Boyle's law and Charles' law each respectively assume what is held constant?
    a) temperature, pressure
    b) temperature, volume
    c) pressure, volume
    d) volume, pressure

    Answer: a Difficulty: 2

FILL-IN-THE-BLANK

60. Two moles of gas are at STP. If the temperature changes to 47.°C and the pressure decreases to half of what it was, what volume does the 2 moles now occupy?

    Answer: 105.liters Difficulty: 2

61. The thermal coefficient of linear expansion of Copper is 1.7×10^{-5} $(C°)^{-1}$. What is the thermal volume expansion coefficient?

    Answer: 5.1×10^{-5} $(C°)^{-1}$ Difficulty: 2

62. A mixture of Helium and another gas diffuses through a porous barrier. The Helium diffuses 3.16 times faster than the other type molecule. Identify the other gas.

    Answer: ARGON: ^{40}Ar Difficulty: 2

Chapter 10

63. Some texts use the kilogram-mole rather than the gram-mole. What would be the values of the gas constant R and Avogadro's number N_A using the kg-mole?

    Answer:
    $R = 8.31 \times 10^3$ J/[(kg-mole) K]
    and $N_A = 6.02 \times 10^{26}$ (kg-mole)$^{-1}$
    Difficulty: 2

64. The RANKINE temperature scale begins at "absolute zero" but uses the Fahrenheit size degree rather than the Celsius degree. Develop a formula to convert C, F, and K (Celsius, Fahrenheit, Kelvin respectively) to R (Rankin).

    Answer: $R = 9/5$ K, $R = 9/5$ C + 491, $R = F + 459$ Difficulty: 2

MULTIPLE CHOICE

65. Diffusion is governed by:
    a) Charles' law
    b) Boyle's law
    c) Archimedes' law
    d) Graham's law
    e) the ideal gas law

    Answer: d Difficulty: 1

FILL-IN-THE-BLANK

66. What is the rms speed of a Helium atom
    a) at 5.°K
    b) at -196.°C (the boiling point of N_2)
    c) at 100.°C

    Answer: a) 176.m/s   b) 692.m/s   c) 1.52×10^3 m/s Difficulty: 2

67. Using K=3/2 k T , what is the average kinetic energy of a He atom at
    a) 10.K
    b) -196.°C
    c) 100.°C

    Answer:
    a) 2.1×10^{-22} J = 1.3 meV   b) 1.59×10^{-21} J = 0.0099 eV
    c) 7.72×10^{-21} J = 0.0482 eV
    Difficulty: 2

MULTIPLE CHOICE

68. A confined gas has an average energy/molecule of 0.06 eV. If the pressure is tripled and the volume is reduced by a factor of 3, what now is the average energy?
   a) 0.02 eV
   b) 0.03 eV
   c) 0.06 eV
   d) 0.10 eV
   e) 0.18 eV

   Answer: c   Difficulty: 2

FILL-IN-THE-BLANK

69. Consider a glass rod encircling the Earth (radius 6378.km). If the linear thermal expansion coefficient of the glass is 3.0×10^{-6} $(C°)^{-1}$, and if the temperature of the rod increases by 1.0 C°, how high off the ground would the expanded ring be (assuming uniform space all around)?

   Answer: 19.meters   Difficulty: 2

ESSAY

70. Describe the anomalous thermal expansion property of water and its significance.

   Answer:
   Near freezing, water has a negative expansion coefficient which means it expands as it cools and freezes. Since ice is less dense than the liquid water, it floats and stays on top of lakes and oceans. This has allowed aquatic life to survive UNDER the ice even during very cold times.

   Difficulty: 2

MULTIPLE CHOICE

71. In the equation $PV = NkT$, the k is known as
   a) Planck's constant
   b) Boltzmann's constant
   c) Avogadro's number
   d) the spring (compressibility) constant

   Answer: b   Difficulty: 1

Chapter 10

TRUE/FALSE

72. Since atomic motion increases with temperature, we expect all materials to expand with an increase of temperature.

    Answer: F Difficulty: 1

MULTIPLE CHOICE

73. At which of the following temperatures would we expect gas molecules to have half the energy that they have at 100.°C ?
    a) 50.°C
    b) -56.°C
    c) -86.°C
    d) 323.K
    e) 93.K

    Answer: c Difficulty: 2

ESSAY

74. A sound wave (pressure disturbance) in air is carried along by the speed of the moving molecule. Since kinetic theory says the "average" speed at room temperature is about 510.m/s why is the speed of sound considerably less (about 340.m/s) than the molecular speed?

    Answer:
    The molecules are moving in random directions. An excited (disturbed) molecule sometimes moves in the direction of the sound wave, at other times perpendicular to it, so it averages out to be a fraction of the maximum speed in the propagation direction.

    Difficulty: 2

Chapter 11

MATCHING

Match the unit to the physical quantity.

a) °C/m
b) W/m² -K⁴
c) °C
d) kcal/kg
e) W/m²
f) kcal/kg-°C
g) dimensionless
h) Btu
i) kcal/m-s-°C

1. thermal gradient

   Answer: a) °C/m   Difficulty: 2

2. Stefan-Boltzmann constant

   Answer: b) W/m²-K⁴   Difficulty: 2

3. dew point

   Answer: c) °C   Difficulty: 2

4. heat of combustion

   Answer: d) kcal/kg   Difficulty: 2

5. intensity of radiation

   Answer: e) W/m²   Difficulty: 2

6. specific heat capacity

   Answer: f) kcal/kg-°C   Difficulty: 2

7. emissivity

   Answer: g) dimensionless   Difficulty: 2

8. English unit of heat

   Answer: h) Btu   Difficulty: 2

9. thermal conductivity

   Answer: i) kcal/m-s-°C   Difficulty: 2

Chapter 11

MULTIPLE CHOICE

10. The mechanical equivalent of heat was first determined by
    a) Lord Kelvin.
    b) Stefan and Boltzmann.
    c) James Joule.
    d) Count Rumford.
    e) Con Edison.

    Answer: d Difficulty: 2

TRUE/FALSE

11. The units for temperature can be converted into units for heat.

    Answer: F Difficulty: 2

MULTIPLE CHOICE

12. A cup of water is scooped up from a swimming pool of water. Compare the temperature T and the internal energy U of the water, in both the cup and the swimming pool.
    a) T_{Pool} is greater than T_{Cup}, and the U is the same.
    b) T_{Pool} is less than T_{Cup}, and the U is the same.
    c) T_{Pool} is equal to T_{Cup}, and U_{Pool} is greater than U_{Cup}.
    d) T_{Pool} is equal to T_{Cup}, and U_{Pool} is less than U_{Cup}.

    Answer: c Difficulty: 2

13. Which of the following is the smallest unit of heat energy?
    a) Calorie
    b) Kilocalorie
    c) Btu
    d) Joule

    Answer: d Difficulty: 2

ESSAY

14. Gasoline yields 4.8×10^7 joules per kg when burned. The density of gasoline is approximately the same as that of water, and 1 gal = 3.8 L. How much energy does your car use on a trip of 100 mi if you get 25 mi per gallon?

    Answer: 0.3×10^8 J Difficulty: 2

MULTIPLE CHOICE

15. What is the power output of a hot air furnace that produces heat at the rate of 160,000 BTU/hr?
    a) 1.6 kW
    b) 22 kW
    c) 47 kW
    d) 80 kW

    Answer: c Difficulty: 2

ESSAY

16. The water flowing over Niagara Falls drops a distance of 50 m. Assuming that all the gravitational energy is converted to thermal energy, by what temperature does the water rise?

    Answer: 0.12° C Difficulty: 2

17. A person tries to heat up her bath water by adding 5 L of water at 80° C to 60 L of water at 30° C. What is the final temperature of the water?

    Answer: 34° C Difficulty: 2

18. A 200-L electric water heater uses 2 kW. Assuming no heat loss, how long would it take to heat water in this tank from 23° C to 75° C?

    Answer: 2170 s Difficulty: 2

19. In grinding a steel knife blade (specific heat = 0.11 cal/g-°C), the metal can get as hot as 400° C. If the blade has a mass of 80 g, what is the minimum amount of water needed at 20° C if the water is not to rise above the boiling point when the hot blade is quenched in it?

    Answer: 33 g Difficulty: 2

MULTIPLE CHOICE

20. It is a well-known fact that water has a higher specific heat capacity than iron. Now, consider equal masses of water and iron that are initially in thermal equilibrium. The same amount of heat, 30 calories, is added to each. Which statement is true?
    a) They remain in thermal equilibrium.
    b) They are no longer in thermal equilibrium; the iron is warmer.
    c) They are no longer in thermal equilibrium; the water is warmer.
    d) It is impossible to say without knowing the exact mass involved and the exact specific heat capacities.

    Answer: b Difficulty: 2

Chapter 11

21. Two equal mass objects (which are in thermal contact) make up a system that is thermally isolated from its surroundings. One object has an initial temperature of 100°C and the other has an initial temperature of 0°C. What is the equilibrium temperature of the system, assuming that no phase changes take place for either object? (The hot object has a specific heat capacity that is three times that of the cold object.)
   a) 25°C
   b) 50°C
   c) 75°C
   d) None of the above

   Answer: c Difficulty: 2

22. A chunk of ice (T = -20°C) is added to a thermally insulated container of cold water (T = 0°C). What happens in the container?
   a) The ice melts until thermal equilibrium is established.
   b) The water cools down until thermal equilibrium is established.
   c) Some of the water freezes and the chunk of ice gets larger.
   d) None of the above.

   Answer: c Difficulty: 2

TRUE/FALSE

23. Steam at 100°C will burn your hand just as severely as water at 100°C will.

   Answer: F Difficulty: 2

MULTIPLE CHOICE

24. Eight grams of water initially at 100°C are poured into a cavity in a very large block of ice initially at 0°C. How many g of ice melt before thermal equilibrium is attained?
   a) 100 g
   b) 10 g
   c) 1 g
   d) An unknown amount; it cannot be calculated without first knowing the mass of the block of ice.

   Answer: b Difficulty: 2

25. A thermally isolated system is made up of a hot piece of aluminum and a cold piece of copper; the aluminum and the copper are in thermal contact. The specific heat capacity of aluminum is more than double that of copper. Which object experiences the greater temperature change during the time the system takes to reach thermal equilibrium?
    a) The aluminum
    b) The copper
    c) Neither; both experience the same size temperature change.
    d) It is impossible to tell without knowing the masses.

    Answer: d Difficulty: 2

26. A thermally isolated system is made up of a hot piece of aluminum and a cold piece of copper; the aluminum and the copper are in thermal contact. The specific heat capacity of aluminum is more than double that of copper. Which object experiences the greater magnitude gain or loss of heat during the time the system takes to reach thermal equilibrium?
    a) The aluminum
    b) The copper
    c) Neither; both experience the same size gain or loss of heat.
    d) It is impossible to tell without knowing the masses.

    Answer: c Difficulty: 2

27. An aluminum kettle (mass = 100 g) of pure water is at room temperature (20°C). The kettle and its contents are placed on a 1000-Watt electric burner and heated to boiling. Assuming that all the heat from the burner heats the kettle and its contents and that a negligible amount of water evaporates before boiling begins, calculate the amount of time required to bring the water to a boil.
    a) 3.5 min
    b) 4.0 min
    c) 7.3 min
    d) 8.1 min

    Answer: a Difficulty: 2

28. Phase changes occur
    a) as the temperature decreases.
    b) as the temperature increases.
    c) as the temperature remains the same.
    d) all of the above are possible.

    Answer: c Difficulty: 2

Chapter 11

29. A block of ice at 0°C is added to a 150-g aluminum calorimeter cup that holds 200 g of water at 10°C. If all but 2 g of ice melt, what was the original mass of the block of ice?
   a) 31.1
   b) 38.8
   c) 42.0
   d) 47.6

   Answer: a   Difficulty: 2

ESSAY

30. The heat of fusion of lead is 5.9 kcal/kg, and the heat of vaporization is 207 cal/kg, and its melting point is 328°C. How much heat is required to melt 50 g of lead initially at 23°C? (The specific heat of lead is 0.031 kcal/kg-°C.)

   Answer: 768 cal   Difficulty: 2

31. The heat of fusion of ice is 80 kcal/kg-°C. When 50 g of ice at 0°C is added to 50 g of water at 25°C, what is the final temperature?

   Answer: 0°C   Difficulty: 2

MULTIPLE CHOICE

32. Turning up the flame under a pan of boiling water causes
   a) the water to boil away faster.
   b) the temperature of the boiling water to increase.
   c) both the water to boil away faster and the temperature of the boiling water to increase.
   d) none of the above.

   Answer: a   Difficulty: 2

33. If heat is added to a pure substance at a steady rate,
   a) its temperature will begin to rise.
   b) it will eventually melt.
   c) it will eventually boil.
   d) more than one of the above is true.
   e) none of the above is true.

   Answer: e   Difficulty: 2

ESSAY

34. At what rate is the human body radiating energy when it is at 33°C? Take the body surface area to be 1.4 m^2, and approximate the body as a blackbody.

   Answer: 733 W   Difficulty: 2

172

35. In an electric furnace used for refining steel, the temperature is monitored by measuring the radiant power emitted through a small hole of area 0.5 cm². The furnace acts like a blackbody radiator. If it is to be maintained at a temperature of 1650°C, at what level should the power radiated through the hole be maintained?

    Answer: 38.8 W Difficulty: 2

TRUE/FALSE

36. The effect of using a large fan in a closed room will be to lower the air temperature.

    Answer: F Difficulty: 2

ESSAY

37. The thermal conductivity of concrete is 0.8 W/m-°C and the thermal conductivity of wood is 0.1 W/m-°C. How thick would a solid concrete wall have to be in order to have the same rate of flow through it as an 8-cm thick wall made of solid wood? (Assume both walls have the same surface area.)

    Answer: 64 cm Difficulty: 2

MULTIPLE CHOICE

38. If you double the absolute temperature of an object, it will radiate energy
    a) 16 times faster.
    b) 8 times faster.
    c) 4 times faster.
    d) none of the above.

    Answer: a Difficulty: 2

39. Convection can occur
    a) only in solids.
    b) only in liquids.
    c) only in gases.
    d) only in liquids and gases.
    e) in solids, liquids, and gases.

    Answer: d Difficulty: 2

Chapter 11

40. Consider two neighboring rectangular houses built from the same materials. One of the houses has twice the length, width, and height of the other. Under identical climatic conditions, what would be true about the rate that heat would have to be supplied to maintain the same inside temperature on a cold day? Compared to the small house, the larger house would need heat supplied at
   a) twice the rate.
   b) 4 times the rate.
   c) 16 times the rate.
   d) none of the above.

   Answer: b Difficulty: 2

41. By what primary heat transfer mechanism does the sun warm the earth?
   a) Convection
   b) Conduction
   c) Radiation
   d) All of the above in combination

   Answer: a Difficulty: 2

42. By what primary heat transfer mechanism does one end of an iron bar become hot when the other end is placed in a flame?
   a) Natural convection
   b) Conduction
   c) Radiation
   d) Forced convection

   Answer: b Difficulty: 2

43. What temperature exists inside a solar collector (effective collection area of 15 m^2) on a bright sunny day when the outside temperature is +20° C? Assume that the collector is thermally insulated, that the sun irradiates the collector with a power per unit area of 600 W/m^2, and that the collector acts as a perfect blackbody.
   a) 73° C
   b) 93° C
   c) 107° C
   d) 154° C

   Answer: b Difficulty: 2

44. The thermal conductivity of aluminum is twice that of brass. Two rods (one aluminum and the other brass) are joined together end to end in excellent thermal contact. The rods are of equal lengths and radii. The free end of the brass rod is maintained at 0°C and the aluminum's free end is heated to 200°C. If no heat escapes from the sides of the rods, what is the temperature at the interface between the two metals?
a) 76°C
b) 133°C
c) 148°C
d) 155°C

Answer: b Difficulty: 2

45. A layer of insulating material with thermal conductivity K is placed on a layer of another material of thermal conductivity 2K. The layers have equal thickness. What is the effective thermal conductivity of the composite sheet?
a) 3K
b) 1.5K
c) K/3
d) 2K/3

Answer: d Difficulty: 2

46. A spaceship is drifting in an environment where the acceleration of gravity is essentially zero. As the air on one side of the cabin is heated by an electric heater, what is true about the convection currents caused by this heating?
a) The hot air around the heater rises and the cooler air moves in to take its place.
b) The hot air around the heater drops and the cooler air moves in to take its place.
c) The convection currents move about the cabin in a random fashion.
d) There are no convection currents.

Answer: c Difficulty: 2

47. If you double the thickness of a wall built from a homogeneous material, the rate of heat loss for a given temperature difference across the thickness will
a) become one-half its original value.
b) also double.
c) become one-fourth its original value.
d) none of the above.

Answer: a Difficulty: 2

Chapter 11

48. When a vapor condenses
    a) the temperature of the substance increases.
    b) the temperature of the substance decreases.
    c) heat energy leaves the substance.
    d) heat energy enters the substance.

    Answer: c Difficulty: 2

49. Frost is an example of
    a) sublimation.
    b) deposition.
    c) condensation.
    d) freezing.

    Answer: b Difficulty: 2

50. In a liquid at a given temperature, the molecules are moving in every direction, some fast, some slowly. Electrical forces of adhesion tend to hold them together. However, occasionally one molecule gains enough energy (as a result of collisions) so that it pulls loose from its neighbors and escapes from the liquid. Which of the following can best be understood in terms of this phenomena?
    a) A hot water bottle will do a better job of keeping you warm than will a rock of the same mass heated to the same temperature.
    b) When a large steel suspension bridge is built, gaps are left between the girders.
    c) If snow begins to fall when you are skiing, you will feel colder than you did before it started to snow.
    d) When you step out of a swimming pool and stand in the wind, you will get colder than you would if you stayed out of the wind.
    e) Increasing the atmospheric pressure over a liquid will cause the boiling temperature to decrease.

    Answer: d Difficulty: 2

51. Which of the following best explains why sweating is important to humans in maintaining suitable body temperature?
    a) Moisture on the skin increases thermal conductivity, thereby allowing heat to flow out of the body more effectively.
    b) Evaporation of moisture from the skin extracts heat from the body.
    c) The high specific heat of water on the skin absorbs heat from the body.
    d) Functioning of the sweat gland absorbs energy that otherwise would go into heating the body.
    e) None of the above explains the principle on which sweating depends.

    Answer: b Difficulty: 2

TRUE/FALSE

52. When going mountain-camping, it will take you less time to boil water than it will at lower altitudes.

    Answer: F Difficulty: 2

53. In a cold climate, its generally better to wear several layers of clothing than one piece of clothing of the same thickness because air has a smaller thermal conductivity than clothing does.

    Answer: T Difficulty: 2

MULTIPLE CHOICE

54. Supersaturation occurs in air when the
    a) relative humidity is 100% and the temperature increases.
    b) relative humidity is less than 100% and the temperature increases.
    c) relative humidity is less 100% and the temperature decreases.
    d) relative humidity is 100% and the temperature decreases.

    Answer: d Difficulty: 2

55. In the vacuum of space, one gets rid of unwanted heat from spacecraft by utilizing:
    a) conduction
    b) convection
    c) radiation
    d) evaporation

    Answer: c Difficulty: 1

FILL-IN-THE-BLANK

56. A bulb delivers 20.watts when its filament is at 1727.°C. If the temperature increases to 227.°C, what power is radiated if the emissivity remains constant?

    Answer: 24.watts Difficulty: 2

Chapter 11

MULTIPLE CHOICE

57. Which of the following walls (identical areas) would be the best thermal INSULATOR? (thermal conductivities, k, are given in J/m-s-C°)

   a) 1.0 inch thick wood (k=0.12)
   b) 20. feet of iron (k=46.)
   c) 5.0 inches of concrete (k=1.3)
   d) 7.0 inches of solid glass (k=0.84)

   Answer:

   b
   Difficulty: 2

ESSAY

58. Glass has more than three times larger thermal conductivity than air. Yet glass-wool (glass fibers and air) is preferable to insulate wall spaces rather than leaving "empty" air spaces. Why is glass-wool better although it has larger conductivity?

   Answer:
   Convection is very important, and the glass fibers reduce the convecting air currents.
   Difficulty: 3

TRUE/FALSE

59. When a vapor condenses into a liquid, its temperature falls.

   Answer: F Difficulty: 1

FILL-IN-THE-BLANK

60. Air conditioners are sometimes rated in "tons". A "ton" represents the freezing in one day of one ton of water at 0.°C to ice at 0.°C. How many watts equal a "ton"?

   Answer: 3.82 Kw Difficulty: 2

TRUE/FALSE

61. A substance with a greater specific heat ALWAYS has a larger heat capacity.

   Answer: T Difficulty: 2

FILL-IN-THE-BLANK

62. Ice cream, when eaten, gives about 2100.Calories/kg to the body. On the other habd, the body must give back energy to warm it up to body temperature (37°C). Assuming ice cream has similar thermal properties to water:
   a) 1/2 kg produces how much energy?
   b) 1/2 kg absorbs how much energy to reach 37.°C?
   c) What is the net gain or loss (eating 1/2 kg ice cream)?

   Answer:
   a) 1050.Cal=4.40 MJ,   b) 0.37 MJ melting and warming,   c) 4.03MJ net body gain.
   Difficulty: 2

63. Alice notices three stars in the night sky, each with a noticeable different color. Sirius appears white, Betelgeuse appears reddish, and Rigel somewhat blue. Place these in order: hot to cold

   Answer: Rigel, Sirius, Betelgeuse (very hot, hot, less hot respectively)

   Difficulty: 2

MULTIPLE CHOICE

64. Most stars radiate much like black bodies. Consider two stars having the same surface temperature, the second having 10 times the diameter of the first. The power radiated per unit area is:

   a) 10 times larger for the 2nd
   b) 100 times larger for the 2nd
   c) the same for both
   d) 10^4 times larger for the 2nd

   Answer: c   Difficulty: 1

65. The point above which no amount of pressure will liquify a gas is called the:
   a) triple point
   b) vaporization point
   c) significant point
   d) critical point

   Answer: d   Difficulty: 2

Chapter 11

FILL-IN-THE-BLANK

66. Name FOUR different units commonly used to quantify heat and sort them largest to smallest.

    Answer: kW-h, BTU, calorie, Joule   Difficulty: 2

TRUE/FALSE

67. A substance with a large specific heat will always have a large latent heat.

    Answer: F Difficulty: 2

MULTIPLE CHOICE

68. What combination of material properties is a measure (indicator) of the insulation quality of a slab of building material?
    a) k   {thermal conductivity
    b) k/d   {conductivity divided by thickness
    c) 1/k   {reciprocal of conductivity
    d) d/k   {thickness divided by conductivity
    e) 1/(dk) {reciprocal of thickness conductivity product

    Answer: d Difficulty: 2

69. The ratio of the partial pressure of water vapor to the saturated water vapor pressure at a given temperature is called the:
    a) dew point
    b) relative humidity
    c) Stefan's law
    d) BTU
    e) latent heat

    Answer: b Difficulty: 1

ESSAY

70. Why do you suppose water has been such a popular heat transfer material (steam heat, hot water systems, etc)?

    Answer:
    It is readily available, but more importantly, it has a relatively high heat capacity and latent heats. Hence it can store large quantities of heat to be transferred elsewhere.
    Difficulty: 2

MULTIPLE CHOICE

71. Which of the following would be the best radiator of thermal energy?
    a) a shiny surface
    b) a perfectly white surface
    c) a perfectly black surface
    d) a metallic surface

    Answer: c Difficulty: 2

ESSAY

72. Why does it take longer to boil an egg in Denver, Colorado, than in Atlanta, Georgia, although the egg is in boiling water at both locations?

    Answer:
    The boiling point (temperature) lowers at the lower pressure at higher elevations. Denver is almost a mile higher than Atlanta and the boiling point there is much lower than in the East.

    Difficulty: 2

MULTIPLE CHOICE

73. At the dew point, the relative humidity is
    a) 0%
    b) 50%
    c) 100%
    d) none of these

    Answer: c Difficulty: 1

74. Which of the following represents the "Mechanical Equivalent of Heat"?
    a) 1 cal = 3.97×10^{-3} BTU
    b) 1 ft-lb = 1.36 Joules
    c) 1 eV = 1.6×10^{-19} J
    d) 1.00 J = 0.239 cal

    Answer: d Difficulty: 2

Chapter 11

ESSAY

75. It was once thought that a "black hole" absorbed all the radiation falling upon it and radiated none. If this were true, what would one infer the temperature of a black hole to be? Actually Stephen Hawking discovered that black holes have a VERY small probability of radiating. Modifying your first answer, would you now infer a large or small tenperature for a B.H.?

Answer:
If no radiation escaped, then its equivalent temperature would be 0.K. But since it radiates (very little) it must be slightly above absolute zero.

Difficulty: 2

Chapter 12

MATCHING

Match the formula to the physical quantity.

a) Q/T
b) Q/m
c) $1 - Q_{cold}/Q_{hot}$
d) Q_{out}/W_{in}
e) $1 - T_{cold}/T_{hot}$
f) $Q - W$
g) $p \Delta V$

1. entropy

    Answer: a) Q/T Difficulty: 2

2. latent heat

    Answer: b) Q/m Difficulty: 2

3. thermal efficiency (non-Carnot)

    Answer: c) $1 - Q_{cold}/Q_{hot}$ Difficulty: 2

4. coefficient of performance

    Answer: d) Q_{out}/W_{in} Difficulty: 2

5. Carnot efficiency

    Answer: e) $1 - T_{cold}/T_{hot}$ Difficulty: 2

6. change in internal energy

    Answer: f) $Q - W$ Difficulty: 2

7. isobaric work

    Answer: g) $p \Delta V$ Difficulty: 2

TRUE/FALSE

8. An example of a reversible process is driving a car.

    Answer: F Difficulty: 2

Chapter 12

MULTIPLE CHOICE

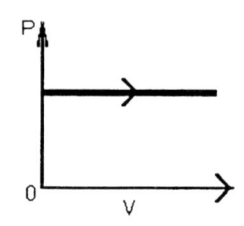

9. The process shown on the p-V graph is an
   a) adiabatic expansion.
   b) isothermal expansion.
   c) isometric expansion.
   d) isobaric expansion.

   Answer: d Difficulty: 2

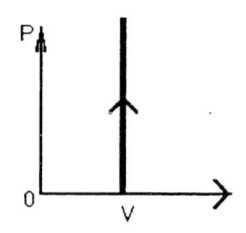

10. The process shown on the p-V graph is an
    a) adiabat.
    b) isotherm.
    c) isomet.
    d) isobar.

    Answer: c Difficulty: 2

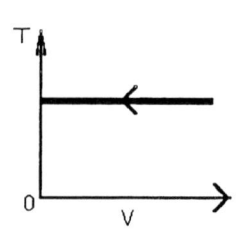

11. The process shown on the T-V graph is an
    a) adiabatic compression.
    b) isothermal compression.
    c) isometric compression.
    d) isobaric compression.

    Answer: b Difficulty: 2

TRUE/FALSE

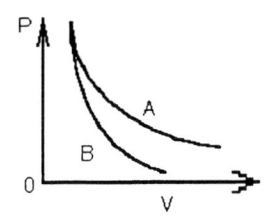

12. (Two processes are shown on the p-V graph; one is an adiabat and the other is an isotherm.) The process represented by the upper curve is an isotherm.

    Answer: T Difficulty: 2

MULTIPLE CHOICE

13. A gas is allowed to expand at constant pressure as heat is added to it. This process is
    a) isothermal.
    b) isometric.
    c) isobaric.
    d) adiabatic.

    Answer: c Difficulty: 2

14. A gas is confined to a rigid container that cannot expand as heat energy is added to it. This process is
    a) isothermal.
    b) isometric.
    c) isobaric.
    d) adiabatic.

    Answer: b Difficulty: 2

15. A gas is expanded to twice its original volume with no change in its temperature. This process is
    a) isothermal.
    b) isometric.
    c) isobaric.
    d) adiabatic.

    Answer: a Difficulty: 2

16. A gas is quickly compressed in an isolated environment. During the event, the gas exchanged no heat with its surroundings. This process is
    a) isothermal.
    b) isometric.
    c) isobaric.
    d) adiabatic.

    Answer: d Difficulty: 2

Chapter 12

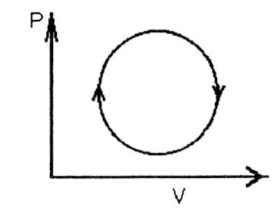

17. An ideal gas is subjected to one complete cycle of the reversible process shown on the p-V graph. The net work done during this cycle is
   a) positive.
   b) negative.
   c) zero.

   Answer: c   Difficulty: 2

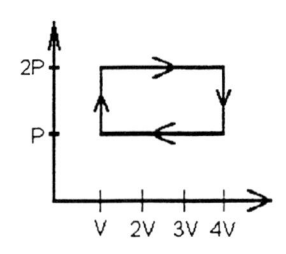

18. A gas is taken through the cycle illustrated here. During one cycle, how much work is done by an engine operating on this cycle?
   a) pV
   b) 2pV
   c) 3pV
   d) 4pV

   Answer: c   Difficulty: 2

ESSAY

19. 200 J of work is done in compressing a gas adiabatically. What is the change in internal energy of the gas?

   Answer: 2000 J   Difficulty: 2

MULTIPLE CHOICE

20. An ideal gas is compressed isothermally from 30 L to 20 L. During this process, 6 J of energy is expended by the external mechanism that compressed the gas. What is the change of internal energy for this gas?
   a) +6 J
   b) zero
   c) −6 J
   d) None of the above

   Answer: b   Difficulty: 2

21. When the first law of thermodynamics, $Q = \Delta U + W$, is applied to an ideal gas that is taken through an isothermal process,
    a) $\Delta U = 0$
    b) $W = 0$
    c) $Q = 0$
    d) none of the above is true

    Answer: a Difficulty: 2

22. When the first law of thermodynamics, $Q = \Delta U + W$, is applied to an ideal gas that is taken through an isobaric process,
    a) $\Delta U = 0$
    b) $W = 0$
    c) $Q = 0$
    d) none of the above is true.

    Answer: d Difficulty: 2

23. When the first law of thermodynamics, $Q = \Delta U + W$, is applied to an ideal gas that is taken through an adiabatic process,
    a) $\Delta U = 0$
    b) $W = 0$
    c) $Q = 0$
    d) none of the above is true.

    Answer: c Difficulty: 2

24. An ideal gas is compressed to one-half its original volume during an isothermal process. The final pressure of the gas
    a) increases to twice its original value.
    b) increases to less than twice its original value.
    c) increases to more than twice its original value.
    d) does not change.

    Answer: a Difficulty: 2

25. A monatomic ideal gas is compressed to one-half its original volume during an adiabatic process. The final pressure of the gas
    a) increases to twice its original value.
    b) increases to less than twice its original value.
    c) increases to more than twice its original value.
    d) does not change.

    Answer: c Difficulty: 2

Chapter 12

26. Consider two cylinders of gas identical in all respects except that one contains O_2 and the other He. Both hold the same volume of gas at STP and are closed by a movable piston at one end. Both gases are now compressed adiabatically to one-third their original volume. Which gas will show the greater temperature increase?
    a) The O_2
    b) The He
    c) Neither; both will show the same increase.
    d) It's impossible to tell from the information given.

    Answer: b Difficulty: 2

27. Consider two cylinders of gas identical in all respects except that one contains O_2 and the other He. Both hold the same volume of gas at STP and are closed by a movable piston at one end. Both gases are now compressed adiabatically to one-third their original volume. Which gas will show the greater pressure increase?
    a) The O_2
    b) The He
    c) Neither; both will show the same increase.
    d) Its impossible to tell form the information given.

    Answer: b Difficulty: 2

ESSAY

28. A container of ideal gas at STP undergoes an isothermal expansion and its entropy changes by 3.66 J/°K. How much work does it do?

    Answer: 999 J Difficulty: 2

29. What is the change in entropy when 50 g of ice melt at 0°C?

    Answer: 4 kcal/°K Difficulty: 2

30. A piece of metal at 80°C is placed in 1.2 L of water at 72°C. The system is thermally isolated and reaches a final temperature of 75°C. Estimate the approximate change in entropy for this process.

    Answer: 2.53 cal/°K Difficulty: 2

MULTIPLE CHOICE

31. When water freezes, the entropy of the water
    a) increases.
    b) decreases.
    c) does not change.
    d) could either increase or decrease; it depends on other factors.

    Answer: b Difficulty: 2

32. Is it possible to transfer heat from a hot reservoir to a cold reservoir?
    a) No!
    b) Yes; this will happen naturally.
    c) Yes, but work will have to be done.
    d) Theoretically yes, but it hasn't been accomplished yet.

    Answer: b Difficulty: 2

TRUE/FALSE

33. When a gas expands adiabatically, the work done by the gas is zero.

    Answer: F Difficulty: 2

34. When a gas is compressed isothermally, the work done by the gas is negative.

    Answer: T Difficulty: 2

35. When a gas is heated isometrically, the work done by the gas is positive.

    Answer: F Difficulty: 2

MULTIPLE CHOICE

36. An ideal gas is expanded isothermally from 20 L to 30 L. During this process, 6 J of energy is expended by the external mechanism that expanded the gas. Which of the following statements is correct?
    a) 6 J of energy flow from surroundings into the gas.
    b) 6 J of energy flow from the gas into the surroundings.
    c) No energy flows into or from the gas since this process is isothermal.
    d) None of the above statements is correct.

    Answer: a Difficulty: 2

37. A certain amount of a monatomic gas is maintained at constant volume as it is cooled by 50°K. This feat is accomplished by removing 400 J of energy from the gas. How much work is done by the gas?
    a) Zero
    b) 400 J
    c) -400 J
    d) None of the above

    Answer: a Difficulty: 2

Chapter 12

38. A monatomic gas is cooled by 50°K at constant volume when 831 J of energy is removed from it. How many moles of gas are in the sample?
   a) 2.50 mol
   b) 1.50 mol
   c) 1.33 mol
   d) None of the above

   Answer: c Difficulty: 2

TRUE/FALSE

39. Freon is commonly used as a refrigerant because it has a relatively high boiling point.

   Answer: F Difficulty: 2

MULTIPLE CHOICE

40. The second law of thermodynamics leads us to conclude that
   a) the total energy of the universe is constant.
   b) disorder in the universe is increasing with the passage of time.
   c) it is theoretically possible to convert heat into work with 100% efficiency.
   d) the average temperature of the universe is increasing with the passage of time.

   Answer: b Difficulty: 2

ESSAY

41. A heat engine absorbs 64 kcal of heat each cycle and exhausts 42 kcal. (a) Calculate the efficiency each cycle. (b) Calculate the work done each cycle.

   Answer: (a) 34%   (b) 22 kcal Difficulty: 2

MULTIPLE CHOICE

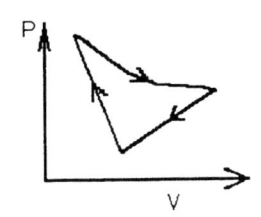

42. A cyclic process is carried out on an ideal gas such that it returns to its initial state at the end of a cycle. If the process was carried out on a clockwise sense around the enclosed area, as shown on the p-V diagram, then that area represents
   a) the heat added to the ideal gas.
   b) the heat that flows from the ideal gas.
   c) the work done by the ideal gas.
   d) the work done on the ideal gas.

   Answer: d Difficulty: 2

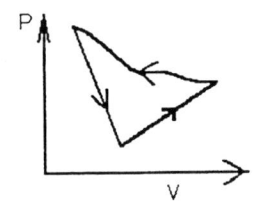

43. A cyclic process is carried out on an ideal gas such that it returns to its initial state at the end of a cycle. If the process was carried out in a counter-clockwise sense around the enclosed area, as shown on the p-V diagram, then that area represents
   a) the heat added to the ideal gas.
   b) the heat that flows from the ideal gas.
   c) the work done by the ideal gas.
   d) the work done on the ideal gas.

   Answer: d Difficulty: 2

Chapter 12

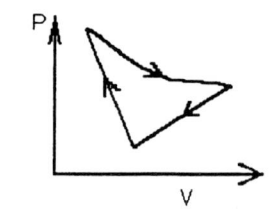

44. A cyclic process is carried out on an ideal gas such that it returns to its initial state at the end of a cycle. If the process was carried out in a clockwise sense around the enclosed area, as shown on the p-V diagram, then the change of internal energy over the full cycle
a) is positive
b) is negative.
c) is zero.
d) cannot be determined from the information given.

Answer: c Difficulty: 2

ESSAY

45. A heat engine operating between 40°C and 380°C has an efficiency 60% of that of a Carnot engine operating between the same temperatures. If the engine absorbs heat at a rate of 60 kW, at what rate does it exhaust heat?

Answer: 41.3 kW Difficulty: 2

46. What is the maximum theoretical efficiency possible for an engine operating between 100°C and 400°C?

Answer: 46% Difficulty: 2

MULTIPLE CHOICE

47. A Carnot cycle consists of
a) two adiabats and two isobars.
b) two isobars and two isotherms.
c) two isotherms and two isomets.
d) two adiabats and two isotherms.

Answer: d Difficulty: 2

48. If the theoretical efficiency of a Carnot engine is to be 100%, the heat sink must be
a) at absolute zero.
b) at 0°C.
c) at 100°C.
d) infinitely hot.

Answer: a Difficulty: 2

49. What is the theoretical efficiency of a Carnot engine that operates between 600° K and 300° K?
   a) 100%
   b) 50%
   c) 25%
   d) None of the above

   Answer: b Difficulty: 2

ESSAY

50. A coal-fired plant generates 600 MW of electric power. The plant uses 4.8×10^6 kg of coal each day. The heat of combustion of coal is 3.3×10^7 J/kg. The steam that drives the turbines is at a temperature of 300° C, and the exhaust water is at 37° C. (a) What is the overall efficiency of the plant for generating electric power? (b) What is the Carnot efficiency? (c) How much thermal energy is wasted each day?

   Answer: (a) 33%   (b) 46%   (c) 1.1×10^{14} J Difficulty: 2

51. One of the most efficient engines built so far has the following characteristics:
   Combustion chamber temperature = 1900° C
   Exhaust temperature = 430° C
   7×10^9 cal of fuel produces 1.4×10^{10} J of work in one hour.
   (a) What is the actual efficiency of this engine? (b) What is the Carnot efficiency of this engine? (c) What is the power output, in hp, of this engine?

   Answer: (a) 48%   (b) 68%   (c) 5213 hp Difficulty: 2

MULTIPLE CHOICE

52. The First Law of Thermodynamics is equivalent to the:
   a) law of conservation of momentum
   b) law of conservation of energy
   c) first law of motion
   d) Newton's third law of motion

   Answer: b Difficulty: 1

Chapter 12

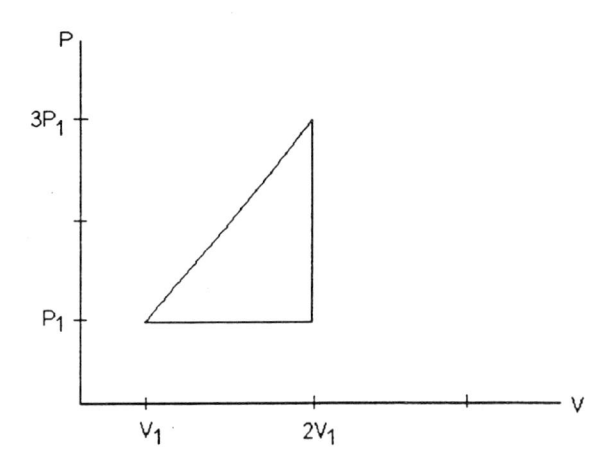

53. A substance is taken through the illustrated cycle from (p_1,v_1) to $(3p_1,2v_1)$ to $(p_1,2v_1)$ back to (p_1,v_1). How much work was done?
a) p_1V_1
b) $2p_1V_1$
c) $3p_1V_1$
d) $4p_1V_1$
e) $5p_1V_1$
f) $6p_1V_1$

Answer: a Difficulty: 2

FILL-IN-THE-BLANK

54. Is it possible for heat to travel from a cold object to a hotter object?

Answer:
Yes (your refrigerator does this all the time) but only with the expenditure of work.

Difficulty: 1

55. What is the third law of thermodynamics?

Answer: It is impossible to reach a temperature of absolute zero.

Difficulty: 1

56. An engine on each cycle takes in 40. Joules, does 10. Joules of work, and expels 30.J of heat. What is its efficiency?

Answer: eff= 25% Difficulty: 2

ESSAY

57. We say that energy can not be created nor destroyed, yet a heat pump usually delivers more energy (heat) into the house than it receives from the power source (electricity, gas, etc). Why is this not a violation of the law of conservation of energy?

    Answer:
    The extra energy was extracted from some environmental source like the outside air or from the ground.
    Difficulty: 2

MULTIPLE CHOICE

58. A reversible engine takes in 40.Joules of heat, does 10.J of work, and expels 30.J of wasted heat during each cycle. If it is operated in reverse as a refrigerator, then its COP would be:
    a) 25%
    b) 4.
    c) 3.
    d) 4%
    e) 3%
    f) 75%

    Answer: c Difficulty: 2

59. The statement that "heat energy cannot be completely transformed into work" is a statement of which thermodynamic law?
    a) zeroth
    b) first
    c) second
    d) third
    e) fourth
    f) fifth

    Answer: c Difficulty: 2

60. As a system loses its ability to do useful work:
    a) energy increases
    b) energy decreases
    c) entropy increases
    d) entropy decreases
    e) none of the above

    Answer: c Difficulty: 1

Chapter 12

ESSAY

61. An inventor tries to sell you his new engine which takes in 40. Joules of heat at 87°C on each cycle, expels 30. Joules at 27°C, and does 10. Joules of work. Why are you not fooled, and can have him prosecuted as a fraud?

    Answer:
    The greatest efficiency possible between those temperatures would be 17% (=1-300/360) and he has claimed a higher efficiency than is physically possible.
    Difficulty: 2

62. Why don't we use "efficiency" for rating refrigerators, like we rate engines, instead of the Coefficient of Performance?

    Answer:
    In both cases we desire a ratio of "what we want" divided by "what it costs"; for a refrig. this is "heat absorbed"/"work input" (COP) but for an engine this is "work out"/"heat in" (=eff)
    Difficulty: 2

MULTIPLE CHOICE

63. The most efficient engine possible is the:
    a) Otto cycle
    b) Kelvin cycle
    c) Wright cycle
    d) Carnot cycle
    e) Joule cycle

    Answer: d Difficulty: 1

64. The change of the internal energy of an ideal gas depends upon:
    a) changing temperature
    b) changing pressure
    c) changing volume
    d) all of the above

    Answer: a Difficulty: 2

FILL-IN-THE-BLANK

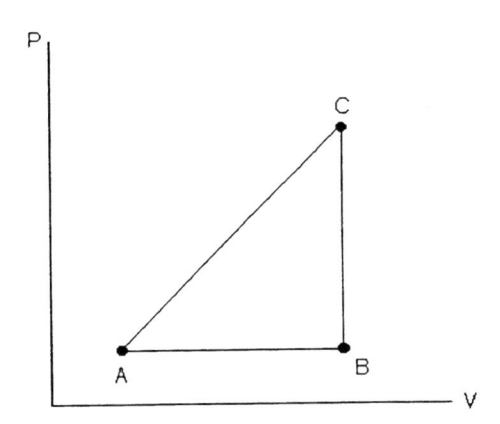

65. Refering to the figure, a substance carried from point A to B absorbs 50.J and finds its internal energy has increased by 20.J. Going from B to C the internal energy decreases by 5.Joules.
    a) how much work was done from A to B?
    b) how much heat was absorbed from B to C?
    c) how much work was done going from B to C?

    Answer: a) 30.J   b) -5.J (i.e. released 5.J)   c) zero

    Difficulty: 2

66. Match up each process with its name:

    ___ at constant temperature   A) isobaric
    ___ at constant volume   B) isochoric
    ___ at constant pressure   C) adiabatic
    ___ no heat added or removed   D) isothermal

    Answer: D , B , A , C   Difficulty: 2

MULTIPLE CHOICE

67. When temperature is plotted against entropy, the area under the process path is equal to:
    a) the work done
    b) the heat transferred
    c) the change in internal energy
    d) the heat added less the work done

    Answer: b   Difficulty: 2

Chapter 12

TRUE/FALSE

68. Real engines have less than the 100% efficiency of the Carnot cycle (the Carnot being the ideal engine).

Answer: F Difficulty: 1

MULTIPLE CHOICE

69. The Otto cycle has how many strokes per cycle during which the volume decreases?
a) 1
b) 2
c) 3
d) 4
e) 8

Answer: b Difficulty: 1

TRUE/FALSE

70. A Carnot cycle requires an ideal gas for its "working substance".

Answer: F Difficulty: 2

FILL-IN-THE-BLANK

71. What is the change in entropy of the lead when 2.0 kg of molten LEAD solidifies. Data: L_V=207.kcal/kg at 1744.°C; L_f=5.9 kcal/kg at 328.°C

Answer: -0.020 kcal/K Difficulty: 2

Chapter 13

MATCHING

Match the unit to the physical quantity.

a) N/m
b) g/cm
c) seconds
d) rad/s
e) degrees
f) Hertz
g) meters

1. spring constant

   Answer: a) N/m Difficulty: 2

2. linear mass density

   Answer: b) g/cm Difficulty: 2

3. period

   Answer: c) seconds Difficulty: 2

4. angular speed

   Answer: d) rad/s Difficulty: 2

5. phase constant

   Answer: e) degrees Difficulty: 2

6. frequency

   Answer: f) Hertz Difficulty: 2

7. wavelength

   Answer: g) meters Difficulty: 2

Chapter 13

MULTIPLE CHOICE

The following question(s) refers to the wave shown below.

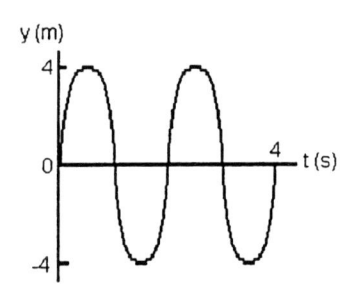

8. The amplitude is
   a) 2 m.
   b) 4 m.
   c) 6 m.
   d) 8 m.

   Answer: b Difficulty: 2

9. The wavelength
   a) is 2 m.
   b) is 4 m.
   c) is 8 m.
   d) cannot be determined from the given information.

   Answer: d Difficulty: 2

10. The frequency is
    a) 0.5 Hz.
    b) 1 Hz.
    c) 2 Hz.
    d) 4 Hz.

    Answer: a Difficulty: 2

ESSAY

11. A pendulum makes 12 complete swings in 8 s. (a) What is its frequency? (b) What is its period?

    Answer: (a) 1.5 Hz   (b) 0.67 s Difficulty: 2

12. A spring-driven dart gun propels a 10-g dart. It is cocked by exerting a force of 20 N over a distance of 5 cm. With what speed will the dart leave the gun, assuming the spring has negligible mass?

    Answer: 14.1 m/s Difficulty: 2

13. A mountain climber of mass 60 kg slips and falls a distance of 4 m, at which time he reaches the end of his elastic safety rope. The rope then stretches an additional 2 m before the climber comes to rest. What is the spring constant of the rope, assuming it obeys Hooke's law?

    Answer: 1760 N/m   Difficulty: 2

14. What is the spring constant of a spring that stretches 2 cm when a mass of 0.6 kg is suspended from it?

    Answer: 294 N/m   Difficulty: 2

MULTIPLE CHOICE

15. What happens to a simple pendulum's frequency if both its length and mass are increased?
    a) It increases.
    b) It decreases.
    c) It remains constant.
    d) It could remain constant, increase, or decrease; it depends on the length to mass ratio.

    Answer: b   Difficulty: 2

16. A simple pendulum consists of a 0.25-kg spherical mass attached to a massless string. When the mass is displaced slightly from its equilibrium position and released, the pendulum swings back and forth with a frequency of 2 Hz. What frequency would have resulted if a 0.50-kg mass (same diameter sphere) had been attached to the string instead?
    a) 1 Hz
    b) 2 Hz
    c) 1.41 Hz
    d) None of the above

    Answer: b   Difficulty: 2

17. The mass of a mass-and-spring system is displaced 10 cm from its equilibrium position and released. A frequency of 4 Hz is observed. What frequency would be observed if the mass had been displaced only 5 cm and then released?
    a) 2 Hz
    b) 4 Hz
    c) 8 Hz
    d) None of the above

    Answer: b   Difficulty: 2

Chapter 13

18. If you take a given pendulum to the moon, where the acceleration of gravity is less than on earth, the resonant frequency of the pendulum will
   a) increase.
   b) decrease.
   c) not change.
   d) either increase or decrease; it depends on its length to mass ratio.

   Answer: b Difficulty: 2

19. A mass is attached to a vertical spring and bobs up and down between points A and B. Where is the mass located when its kinetic energy is a maximum?
   a) At either A or B
   b) Midway between A or B
   c) One-fourth of the way between A and B
   d) None of the above

   Answer: b Difficulty: 2

20. A mass is attached to a vertical spring and bobs up and down between points A and B. Where is the mass located when its kinetic energy is a minimum?
   a) At either A or B
   b) Midway between A or B
   c) One-fourth of the way between A and B
   d) None of the above

   Answer: a Difficulty: 2

21. A mass vibrates back and forth from the free end of an ideal spring (k = 20 N/m) with an amplitude of 0.25 m. What is the maximum kinetic energy of this vibrating mass?
   a) 2.5
   b) 1.25
   c) 0.625
   d) It is impossible to give an answer since kinetic energy cannot be determined without knowing the object's mass.

   Answer: c Difficulty: 2

22. Two masses, A and B, are attached to different springs. Mass A vibrates with an amplitude of 8 cm at a frequency of 10 Hz and mass B vibrates with an amplitude of 5 cm at a frequency of 16 Hz. How does the maximum speed of A compare to the maximum speed of B?
   a) Mass A has the greater maximum speed.
   b) Mass B has the greater maximum speed.
   c) They are equal.

   Answer: c Difficulty: 2

23. A mass vibrates back and forth from the free end of an ideal spring (k = 20 N/m) with an amplitude of 0.30 m. What is the kinetic energy of this vibrating mass when it is 0.30 m from its equilibrium position?
   a) Zero
   b) 0.90
   c) 0.45
   d) It is impossible to give an answer without knowing the object's mass.

   Answer: c Difficulty: 2

TRUE/FALSE

24. When the length of a simple pendulum is tripled, the time for one complete vibration increases by a factor of 3.

   Answer: T Difficulty: 2

MULTIPLE CHOICE

25. When the mass of a simple pendulum is tripled, the time required for one complete vibration
   a) increases by a factor of 3.
   b) does not change.
   c) decreases to one-third of its original value.
   d) decreases to 1/3 of its original value

   Answer: b Difficulty: 2

26. A mass is attached to a spring. It oscillates at a frequency of 4/0 Hz when displaced a distance of 2 cm from equilibrium and released. What is the maximum velocity attained by the mass?
   a) 0.02 m/s
   b) 0.04 m/s
   c) 0.08 m/s
   d) 0.16 m/s
   e) 0.32 m/s

   Answer: d Difficulty: 2

Chapter 13

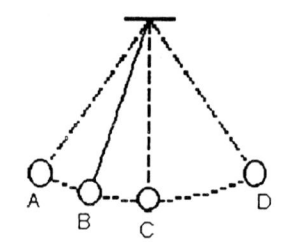

27. A mass on the end of a massless spring undergoes SHM. Where is the instantaneous acceleration of the mass greatest?
    a) A and C
    b) B
    c) C
    d) A and D

    Answer: d Difficulty: 2

28. A simple pendulum consists of a mass M attached to a weightless string of length L. For this system, when undergoing small oscillations
    a) the frequency is proportional to the amplitude.
    b) the period is proportional to the amplitude.
    c) the frequency is independent of the mass M.
    d) the frequency is independent of the length L.

    Answer: c Difficulty: 2

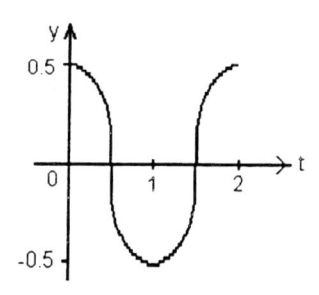

29. The equation of motion of the wave shown is
    a) y = 0.5 sin (0t)
    b) y = 0.5 sin (40t)
    c) y = 0.5 cos (0t)
    d) y = 0.5 cos (40t)
    e) none of the above.

    Answer: c Difficulty: 2

ESSAY

30. A 0.3-kg mass is suspended on a string. In equilibrium the mass stretches the spring 2 cm downward. The mass is then pulled an additional distance of 1 cm down and released from rest. (a) Calculate the period of oscillation. (b) Calculate the total energy of the system. (c) Write down its equation of motion. (d) Calculate the velocity of the mass at T/3 s.

    Answer: (a) 0.0126 cm (b) 0.0074 J (c) y = 0.01 cos(22.1 t) (d) 0.19 m/s

    Difficulty: 2

31. The equation of motion of a particle undergoing simple harmonic motion is y = 2 sin(.6 t), where y is in cm. At t = 0.6 s determine the particle's (a) displacement (b) velocity (c) acceleration.

    Answer: (a) 0.0126 cm   (b) 1.20 cm/s   (c) 0.0045 cm/s^2 Difficulty: 2

32. A simple pendulum has a length of 0.8 m and a bob of mass 0.20 kg. It is initially displaced to an angle of 18° from the vertical and released from rest. At t = 1.2 s, (a) what angle does it make with the vertical? (b) what is its velocity?

    Answer: (a) 8.8°   (b) 0.96 m/s Difficulty: 2

MULTIPLE CHOICE

33. A 2-kg mass is hung from a spring (k = 18 N/m), displaced slightly from its equilibrium position, and released. What is the frequency of its vibration?
    a) 1.5/0 Hz
    b) 3.0/0 Hz
    c) 1.5 Hz
    d) None of the above

    Answer: a Difficulty: 2

34. Increasing the mass m of a mass-and-spring system causes what kind of change on the resonant frequency of the system?
    a) The frequency increases.
    b) The frequency decreases.
    c) There is no change in the frequency.
    d) The frequency increases if the ratio k/m is greater than or equal to 1 and decreases if the ratio k/m is less than 1.

    Answer: b Difficulty: 2

Chapter 13

35. Increasing the spring constant k of a mass-and-spring system causes what kind of change in the resonant frequency of the system? (Assume no change in the system's mass m.)
   a) The frequency increases.
   b) The frequency decreases.
   c) There is no change in the frequency.
   d) The frequency increases if the ratio k/m is greater than or equal to 1 and decreases if the ratio k/m is less than 1.

   Answer: a Difficulty: 2

36. A mass attached to the free end of an ideal spring executes SHM according to the equation x = 0.5 sin(18 t) where x is in meters and t is in seconds. What is the maximum velocity of the mass?
   a) 36 m/s
   b) 3 m/s
   c) 9 m/s
   d) None of the above

   Answer: c Difficulty: 2

37. A mass attached to the free end of an ideal spring executes SHM according to the equation x = 0.5 sin(20 t) where x is in meters and t is in seconds. What is the magnitude of the maximum acceleration for this mass?
   a) 200 m/s^2
   b) 20 m/s^2
   c) 10 m/s^2
   d) None of the above

   Answer: c Difficulty: 2

38. What mass should be attached to a vertical spring (k = 39.5 N/m) so that the natural vibration frequency of the system will be 1.00 Hz?
   a) 39.5 kg
   b) 6.29 kg
   c) 2.00 kg
   d) 1.00 kg

   Answer: a Difficulty: 2

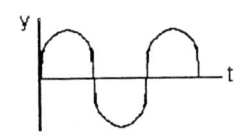

Shown here is a graph of velocity vs. time for a system undergoing simple harmonic motion. Which of the other graphs represents the system's acceleration as a function of time?

39.

Answer: a   Difficulty: 2

40. Simple pendulum A swings back and forth at twice the frequency of simple pendulum B. Which statement is correct?
a) Pendulum B is twice as long as A.
b) Pendulum B is twice as massive as A.
c) The length of B is 1.41 times the length of A.
d) The mass of B is 1.41 times the mass of A.

Answer: c   Difficulty: 2

ESSAY

41. A long rope of linear density 0.25 kg/m is under tension. Transverse waves travel on it at 30 m/s. How much energy is stored in 1 m of the rope when a driving frequency of 6 Hz produces a traveling wave with an amplitude of 0.10 m?

Answer: 0.043 J/m   Difficulty: 2

Chapter 13

MULTIPLE CHOICE

42. A 2-kg mass is attached to the end of a horizontal spring (k = 50 N/m) and set into simple harmonic motion with an amplitude of 0.1 m. What is the total mechanical energy of this system?
    a) 0.02 J
    b) 25 J
    c) 0.25 J
    d) 1.00 J
    e) 2.5 J

    Answer: c Difficulty: 2

43. Doubling only the amplitude of a vibrating mass-and-spring system produces what effect on the system's mechanical energy?
    a) Increases the energy by a factor of two.
    b) Increases the energy by a factor of three.
    c) Increases the energy by a factor of four.
    d) Produces no change.

    Answer: c Difficulty: 2

ESSAY

44. A string of linear density 6 g/m is under a tension of 180 N. What is the velocity of propagation of transverse waves along the string?

    Answer: 173 m/s Difficulty: 2

MULTIPLE CHOICE

45. What is the velocity of propagation if a wave has a frequency of 12 Hz and a wavelength of 3m?
    a) 0.25 m
    b) 4 m
    c) 36 m
    d) None of the above

    Answer: c Difficulty: 2

46. What is the amplitude (in meters) of a wave whose displacement is given by y = 0.5 sin(0.20x + 120t) m?
    a) 0.5 m
    b) 6 m
    c) 10 m
    d) None of the above

    Answer: a Difficulty: 2

47. What is the wavelength (in meters) of a wave whose displacement is given by y = 0.5 sin(0.20x + 120t) m?
   a) 0.5 m
   b) 6 m
   c) 10 m
   d) None of the above

   Answer: c Difficulty: 2

48. What is the frequency (in Hz) of a wave whose displacement is given by y = 0.5 sin(0.20x + 120t) m?
   a) 0.5 Hz
   b) 6 Hz
   c) 10 Hz
   d) None of the above

   Answer: b Difficulty: 2

49. The lower the frequency of a sound wave, the
   a) lower its velocity.
   b) greater its wavelength.
   c) smaller its amplitude.
   d) shorter its period.

   Answer: b Difficulty: 2

ESSAY

50. Find the first three harmonics of a string of linear mass density 2 g/m and length 0.60 m when it is subjected to tension of 50 N.

    Answer: 132 Hz, 264 Hz, 396 Hz Difficulty: 2

51. A stretched string is observed to have three equal segments in a standing wave driven at a frequency of 480 Hz. What driving frequency will set up a standing wave with four equal segments?

    Answer: 640 Hz Difficulty: 2

MULTIPLE CHOICE

52. Consider a traveling wave on a string of length L, mass M, and tension T. A standing wave is set up. Which of the following is true?
   a) The wave velocity depends on M, L, T.
   b) The wavelength of the wave is proportional to the frequency.
   c) The particle velocity is equal to the wave velocity.
   d) The wavelength is proportional to L.

   Answer: a Difficulty: 2

Chapter 13

53. A string of linear density 1.5 g/m is under a tension of 20 N. What should its length be if its fundamental resonance frequency is 220 Hz?
   a) 0.85 m
   b) 0.96 m
   c) 1.05 m
   d) 1.12 m

   Answer: a Difficulty: 2

54. The velocity of propagation of a transverse wave on a 2-m long string fixed at both ends is 200 m/s. Which one of the following is not a resonant frequency of this string?
   a) 25 Hz
   b) 50 Hz
   c) 100 Hz
   d) 200 Hz

   Answer: a Difficulty: 2

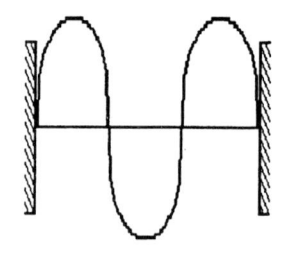

55. A spring, fixed at both ends, vibrates at a frequency of 12 Hz with a standing transverse wave pattern as shown. What is this spring's fundamental frequency?
   a) 36 Hz
   b) 16 Hz
   c) 8 Hz
   d) 4 Hz

   Answer: d Difficulty: 2

56. If one doubles the tension in a violin string, the fundamental frequency of that string will be how many times the original frequency?
   a) 2
   b) 4
   c) 1.41
   d) none of these

   Answer: c Difficulty: 2

57. The total distance traveled by an object in one complete cycle of Simple Harmonic Motion is ____ times the amplitude.
    a) half
    b) one
    c) two
    d) three
    e) four

    Answer: e Difficulty: 2

FILL-IN-THE-BLANK

58. Tarzan swings back and forth on a long vine. His friend Jane notices 30. swings in 2.4 minutes.

    a) The frequency of the swing is _____

    b) The vine length is _____ meters.

    Answer: a) 0.21 Hertz,   b) 5.7 m   Difficulty: 2

MULTIPLE CHOICE

59. Waves on a lake pass under a floating bird causing the bird to bob up and down with a period of 2.5 seconds. If the distance from wave trough to wave trough is 3.0 meters, what is the speed of the wave?
    a) 1.5 m/s
    b) 0.21 m/s
    c) 1.88 m/s
    d) 1.2 m/s
    e) 2.1 m/s

    Answer: d Difficulty: 2

60. The total energy stored in simple harmonic motion is proportional to:
    a) the amplitude
    b) reciprocal of the spring constant
    c) square of the spring constant
    d) square of the amplitude

    Answer: d Difficulty: 1

61. Simple Harmonic Motion is characterized by:
    a) constant acceleration
    b) acceleration proportional to velocity
    c) acceleration proportional to displacement
    d) acceleration proportional to the acceleration of gravity

    Answer: c Difficulty: 1

Chapter 13

62. Grandfather clocks often are built so that each one-way swing of the pendulum is a second. How long is the length of a simple pendulum for a 2.00 second period?
a) 0.993 m
b) 24.8 cm
c) 101.cm
d) 0.500 m

Answer: a Difficulty: 2

FILL-IN-THE-BLANK

63. SHM may be written y = A sin(wt + p). What is the value of p, the phase constant, when:
a) the initial displacement is zero _____
b) the initial displacement is A _____
c) the initial velocity is zero _____

Answer: a) p=0   b) p=π   c) p= +π or -π   Difficulty: 2

64. Name 5 or more different type of waves or wave motion.

Answer:
sound, light, water waves, earthquake waves, waves on a plucked string, ...
Difficulty: 1

ESSAY

65. Why can longitudinal earthquake waves go straight through the center of the Earth but transverse waves cannot?

Answer: The transverse wave can not traverse the outer liquid core.

Difficulty: 2

TRUE/FALSE

66. When waves interfere, the result is always smaller than the original waves ("they interfere").

Answer: F Difficulty: 1

FILL-IN-THE-BLANK

67. Give at least one example of the following:
    a) longitudinal standing wave.
    b) transverse standing wave.

    Answer:
    a) sound wave resonating in an organ pipe, b) vibrating string on a violin.
    Difficulty: 2

MULTIPLE CHOICE

68. Compare two vibrating strings, each has the same tension, but the second has four times the mass density of the first. The wave speed along the 2nd string is how many times the speed along the 1st?
    a) 1/16
    b) 1/4
    c) 1/2
    d) 2
    e) 4
    f) 16

    Answer: c   Difficulty: 2

FILL-IN-THE-BLANK

69. What is the equation of motion of a 30.gram mass on a spring of stiffness 3.0 N/m if it is initially displaced 7.7 cm and released?

    Answer: y = 7.7 cos(10 t)   where y is in cm and t in seconds.

    Difficulty: 2

70. If a floating log is seen to bob up and down 15 times in a minute as waves pass beneath. What is the frequency and period of the wave?

    Answer: frequency = 0.25 Hz , period = 4.0 seconds   Difficulty: 2

71. Does Simple Harmonic Motion occur when the force, acting on a mass, is proportional to the displacement? (explain)

    Answer:
    False. F=kx (k positive) leads to unconstrained motion. One must have the force in opposite direction: F=-kx to produce SHM

    Difficulty: 2

Chapter 13

ESSAY

72. How would you "weigh" the astronauts in orbit (where they feel weightless) so that you can keep them in good health?

    Answer:
    You could determine their MASS by harnessing them to springs, setting them to oscillate (SHM) and timing their period of oscillation. Knowing the "spring constant" allows one to calculate the mass. [Remember $w^2 = k/m$]

    Difficulty: 2

73. Imagine hitting a heavy anvil with a hammer. The hammer is in contact with the metal for a short period of time. How do you suppose the time of contact depend upon how hard you hit the anvil? i.e. Does a hard hit remain in contact longer or shorter than a light tap?
    [It is reasonable to assume Hooke's law of elasticity to hold]

    Answer:
    The time of contact does NOT depend upon the strength of the blow. The PERIOD of oscillation (the time in contact is half a period!) is independent of the amplitude for SHM.

    Difficulty: 2

FILL-IN-THE-BLANK

74. Spock has landed on a newly discovered planet and is instructed to determine its gravitational strength. He constructs a simple pendulum with a length of 700.mm and observes 20. swings in 1 minute and 16.7 seconds. What does he deduce the "acceleration of gravity" to be from this ?

    Answer: $g = 1.88$ m/s^2 Difficulty: 2

TRUE/FALSE

75. In a vibrating string, the antinodes experience the maximum acceleration.

    Answer: T Difficulty: 2

MULTIPLE CHOICE

76. Suppose you wish to lower the pitch of a violin string by 2 octaves. This could be done by:
    a) increasing the string tension a factor of four.
    b) reducing the string tension by a factor of four.
    c) replacing the string with one of the same material only half as massive.
    d) replacing the string with one of the same material but with quadruple the diameter.
    e) shortening the length to a quarter as much.

    Answer: d Difficulty: 2

Chapter 14

MATCHING

Match the unit to the physical quantity.

a) J/s
b) decibel
c) dimensionless
d) W/m²
e) meters
f) Hertz
g) degrees

1. power

   Answer: a) J/s Difficulty: 2

2. intensity level

   Answer: b) decibel Difficulty: 2

3. Mach number

   Answer: c) dimensionless Difficulty: 2

4. intensity

   Answer: d) W/m² Difficulty: 2

5. path difference

   Answer: e) meters Difficulty: 2

6. beat frequency

   Answer: f) Hertz Difficulty: 2

7. phase difference

   Answer: g) degrees Difficulty: 2

ESSAY

8. What is the ratio of the speed of sound in air at 0°C to the speed at 100°C?

   Answer: 0.86 Difficulty: 2

9. On a day when the speed of sound in air is 340 m/s, a bat emits a shriek whose echo reaches it 0.025 s later. How far away was the object that reflected back the sound?

   Answer: 4.25 m Difficulty: 2

10. Sound traveling in air at 23°C enters a cold front where the air temperature is 2°C. If the sound frequency is 1500 Hz, determine the wavelength in the warmer air and in the colder air.

Answer: 345 m/s; 332 m/s   Difficulty: 2

MULTIPLE CHOICE

11. If you hear thunder 5 s after seeing a flash of lightning, the distance to the lightning strike is about
    a) 600 m
    b) 1200 m
    c) 1700 m
    d) 2200 m

    Answer: c   Difficulty: 2

12. When sound passes from air into water
    a) its wavelength does not change.
    b) its frequency does not change.
    c) its velocity does not change.
    d) all of the above are true.

    Answer: b   Difficulty: 2

13. What happens to the velocity of sound in a gas when the absolute temperature of that gas is doubled?
    a) It doubles.
    b) It quadruples.
    c) It increases to 1.41 times its original value.
    d) None of the above.

    Answer: c   Difficulty: 2

14. Sound vibrations with frequency less than 20 Hz are called
    a) infrasonics.
    b) ultrasonics.
    c) supersonics.
    d) none of the above.

    Answer: a   Difficulty: 2

15. Sound vibrations with frequencies greater than 20,000 Hz are called
    a) infrasonics.
    b) ultrasonics.
    c) supersonics.
    d) none of the above.

    Answer: b   Difficulty: 2

Chapter 14

16. As the temperature of the air increases, what happens to the velocity of sound? (Assume that all other factors remain constant.)
    a) It increases.
    b) It decreases.
    c) It does not change.
    d) It increases when atmospheric pressure is high and decreases when the pressure is low.

    Answer: a Difficulty: 2

17. As atmospheric pressure increases, what happens to the velocity of sound? (Assume that all other factors remain constant.)
    a) It increases.
    b) It decreases.
    c) It does not change.
    d) It increases on warm days and decreases on cold days.

    Answer: c Difficulty: 2

18. The wavelength in air of a sound wave of frequency 500 Hz is
    a) 0.68 m
    b) 0.75 m
    c) 1.47 m
    d) 1.80 m
    e) 2.00 m

    Answer: a Difficulty: 2

19. Which of the following is a false statement?
    a) Sound waves are longitudinal pressure waves.
    b) Sound can travel through vacuum.
    c) Light travels very much faster than sound.
    d) The transverse waves on a vibrating string are different from sound waves.
    e) "Pitch" (in music) and frequency have approximately the same meaning.

    Answer: b Difficulty: 2

20. If you were to inhale a few breaths from a helium gas balloon, you would probably experience an amusing change in your voice. You would sound like Donald Duck or Alvin the Chipmunk. What is the cause of this curious high-pitched effect?
    a) The helium causes your vocal cords to tighten and vibrate at a higher frequency.
    b) For a given frequency of vibration of your vocal cords, the wavelength of sound is less in helium than it is in air.
    c) Your voice box is resonating at the second harmonic, rather than at the fundamental frequency.
    d) Low frequencies are absorbed in helium gas, leaving the high frequency components, which result in the high, squeaky sound.
    e) Sound travels faster in helium than in air at a given temperature.

    Answer: e Difficulty: 2

Testbank

ESSAY

21. What is the intensity of a 70-dB sound?

    Answer: 10^{-5} W/m² Difficulty: 2

22. What is the ratio of the power levels of two sounds with intensity levels of 40 dB and 70 dB?

    Answer: 1000:1 Difficulty: 2

23. The intensity at a distance of 6 m from a source is 6×10^{-10} W.
    (a) What is the power level emitted by the source? (b) What is the intensity level in dB?

    Answer: (a) 999×10^{-8} W/m² (b) 37 dB Difficulty: 2

MULTIPLE CHOICE

24. What is the intensity level of a sound with intensity 10^{-3} W/m²?
    a) 30 dB
    b) 60 dB
    c) 90 dB
    d) 96 dB

    Answer: c Difficulty: 2

25. You double your distance from a constant sound source that is radiating equally in all directions. What happens to the intensity of the sound? It reduces to
    a) one-half its original value.
    b) one-fourth its original value.
    c) one-sixteenth its original value.
    d) none of the above.

    Answer: b Difficulty: 2

26. You double your distance from a constant sound source that is radiating equally in all directions. What happens to the intensity level of the sound? It drops by
    a) 2 dB.
    b) 3 dB.
    c) 6 dB.
    d) 8 dB.

    Answer: c Difficulty: 2

Chapter 14

TRUE/FALSE

27. At a rock concert, if the speakers are placed on the floor instead of higher up on a tower, the intensity of the sound you would hear, from the same distance away, is the same.

    Answer: F   Difficulty: 2

28. If the phase angle between sound waves from two different point sources is 180°, destructive interference will occur.

    Answer: T   Difficulty: 2

MULTIPLE CHOICE

29. The Doppler shift explains
    a) why the siren on a police car changes its pitch as it races past us.
    b) why a sound grows quieter as we move away from the source.
    c) how sonar works.
    d) the phenomenon of beats.
    e) why it is that our hearing is best near 3000 Hz.

    Answer: a   Difficulty: 2

30. Suppose that a source of sound is emitting waves uniformly in all directions. If you move to a point twice as far away from the source, the frequency of the sound will be
    a) unchanged.
    b) half as great.
    c) one-fourth as great.
    d) twice as great.

    Answer: c   Difficulty: 2

ESSAY

31. Two adjacent sources each emit a frequency of 800 Hz in air where the velocity of sound is 340 m/s. How much farther back would source 1 have to be moved so an observer in front of the sources would hear no sound?

    Answer: 0.213 m   Difficulty: 2

32. What is the beat frequency of two sounds that have equal amplitudes and frequencies of 440 Hz and 444 Hz?

    Answer: 4 Hz   Difficulty: 2

33. The corresponding violin strings on two violins in an orchestra are found to produce a beat frequency of 2 Hz when a frequency of 660 Hz is played. What percentage change in the tension of one of the strings would bring them to the same frequency?

Answer: 0.6% Difficulty: 2

MULTIPLE CHOICE

34. Which of the following increases as a sound becomes louder?
    a) Frequency
    b) Wavelength
    c) Amplitude
    d) Period
    e) Velocity

Answer: c Difficulty: 2

35. Consider the standing wave on a guitar string and the sound wave generated by the guitar as a result of this vibration. What do these two waves have in common?
    a) They have the same wavelength.
    b) They have the same velocity.
    c) They have the same frequency.
    d) More than one of the above is true.
    e) None of the above is true.

Answer: c Difficulty: 2

36. If a jet plane were to double its MACH-speed, its half-angle will decrease by a factor of:
    a) "
    b) 2
    c) the arc-sin (")
    d) none of the above.

Answer: d Difficulty: 2

37. The half angle of the conical shock wave produced by a supersonic aircraft is 60°. What is the Mach number of the aircraft?
    a) 0.87
    b) 1.15
    c) 1.73
    d) 2.0

Answer: b Difficulty: 2

Chapter 14

38. A jet flies at a speed of Mach 1.4. What is the half-angle of the conical shock wave formed?
    a) 30°
    b) 36°
    c) 44°
    d) 46°

    Answer: c Difficulty: 2

39. Two pure tones are sounded together and a particular beat frequency is heard. What happens to the beat frequency if the frequency of one of the tones is increased?
    a) It increases.
    b) It decreases.
    c) It does not change.
    d) It could either increase or decrease.

    Answer: d Difficulty: 2

40. An unknown tuning fork is sounded along with a tuning fork whose frequency is 256 Hz and a beat frequency of 3 Hz is heard. What is the frequency of the unknown tuning fork?
    a) It must be 259 Hz.
    b) It must be 253 Hz.
    c) It could be either 253 Hz or 259 Hz; there is no way to tell.
    d) It must be 256 Hz.

    Answer: c Difficulty: 2

41. In order to produce beats, the two sound waves should have
    a) the same amplitude.
    b) slightly different amplitudes.
    c) the same frequency.
    d) slightly different frequencies.

    Answer: d Difficulty: 2

42. A sound source (normal frequency of 1000 Hz) approaches a stationary observer at one-half the speed of sound. The observer hears a frequency of
    a) 2000 Hz.
    b) 500 Hz.
    c) 1500 Hz.
    d) none of the above.

    Answer: a Difficulty: 2

43. An observer approaches a stationary 1000-Hz sound source at twice the speed of sound. The observer hears a frequency of
   a) 4000 Hz.
   b) 2000 Hz.
   c) 500 Hz.
   d) none of the above.

   Answer: d Difficulty: 2

ESSAY

44. An organ pipe open at both ends has a length of 0.80 m. If the velocity of sound in air is 340 m/s, what are the frequencies of the second and third harmonics?

   Answer: 425 Hz; 638 Hz Difficulty: 2

MULTIPLE CHOICE

45. A closed organ pipe of length 0.75 m is played when the speed of sound in air is 34 m/s. What is the fundamental frequency?
   a) 57 Hz
   b) 114 Hz
   c) 171 Hz
   d) 228 Hz

   Answer: b Difficulty: 2

46. Which of the following properties of a sound wave is most closely identified with the "pitch" of a musical note?
   a) Amplitude
   b) Wavelength
   c) Frequency
   d) Phase

   Answer: c Difficulty: 2

47. Which of the following is most closely identified with loudness of a musical note?
   a) Frequency
   b) Velocity
   c) Phase
   d) Amplitude

   Answer: d Difficulty: 2

TRUE/FALSE

48. Only odd harmonics can be produced in an open organ pipe.

   Answer: F Difficulty: 2

Chapter 14

MULTIPLE CHOICE

49. The lowest tone to resonate in a closed pipe of length L is 200 Hz. Which of the following frequencies will not resonate in that pipe?
    a) 400 Hz
    b) 60 Hz
    c) 1000 Hz
    d) 1400 Hz

    Answer: a Difficulty: 2

50. The lowest tone to resonate in an open pipe of length L is 400 Hz. What is the frequency of the lowest tone that will resonate in an open pipe of length 2L?
    a) 800 Hz
    b) 200 Hz
    c) 1600 Hz
    d) 100 Hz

    Answer: b Difficulty: 2

FILL-IN-THE-BLANK

51. The audible range of sound frequencies is what?

    Answer: 20 Hz to 20 kHz Difficulty: 1

MULTIPLE CHOICE

52. What is the speed of sound in air at freezing temperature?
    a) 331 mph
    b) 650 m/s
    c) 331 m/s
    d) 650 mph
    e) 1100 m/s
    f) 3×10^8 m/s

    Answer: c Difficulty: 1

TRUE/FALSE

53. Since the Helium molecule (He) is 2 times as massive as the Hydrogen molecule (H_2), the speed of sound in Hydrogen is twice as large as the speed in Helium.

    Answer: F Difficulty: 1

MULTIPLE CHOICE

54. If the intensity of sound changes (increases) by a factor of 100, the decibel level changes by a factor of:
    a) 1
    b) 10
    c) 20
    d) 100
    e) log 100

    Answer: c Difficulty: 2

FILL-IN-THE-BLANK

55. The distance from the Moon to Earth is 3.8×10^5 km.
    a) How long does it take a light beam to travel this distance?
    b) If instead of a vacuum, the space between had air at STP, how long would it take for sound to travel that same distance?

    Answer: a) 1.3 seconds for light, b) 1.1×10^6 s = 1.3 days for sound

    Difficulty: 2

56. After seeing a flash of lightning, if one counts the seconds before hearing the flash, one can estimate the distance to the flash. How many seconds delay
    a) per mile?
    b) per kilometer?

    Answer: a) 5. seconds per mile,   b) 3. seconds per km Difficulty: 2

57. A train moving 30.m/s approaches a station and sounds its whistle and passengers standing at the station hear a pitch of 440. cycles/second. What pitch will be heard when the train is at rest at the station? (assume the air temperature is 20° C)

    Answer: 402.Hz Difficulty: 2

58. A train moving 30.m/s approaches a station and sounds its whistle and passengers standing at the station hear a pitch of 440. cycles/second. What pitch will be heard when the train is later moving away from the station at 30.m/s? (assume the air temperature is 20° C)

    Answer: 370.Hz Difficulty: 2

Chapter 14

MULTIPLE CHOICE

59. The pitch of sound is related to its:
    a) loudness
    b) amplitude
    c) velocity
    d) intensity
    e) frequency

    Answer: e Difficulty: 2

60. The decibel level of sound is related to its:
    a) frequency
    b) intensity
    c) velocity
    d) wavelength

    Answer: b Difficulty: 2

FILL-IN-THE-BLANK

61. Estimate by what percent the perceived frequency of speech changes due to a talking person walking away from rest to a speed of 3 m/s.

    Answer: decreases by 1% Difficulty: 2

MULTIPLE CHOICE

62. How many times more intense is an 80 db sound than a 60 db sound?
    a) 2
    b) 10
    c) 20
    d) 1.33
    e) 100
    f) 140

    Answer: e Difficulty: 2

63. A sound wave of pure frequency can under certain circumstances cause a glass goblet to shatter. This would be an example of:
    a) constructive interference
    b) resonance
    c) overtones
    d) destructive interference

    Answer: b Difficulty: 2

64. Two coherent sound wave will destructively interfere when the path difference is the wavelength times what?
a) an integer
b) half an even interger
c) half an odd integer
d) an odd integer

Answer: c Difficulty: 2

65. Two coherent sound waves will constructively interfere when the path difference is the wavelength times what?
a) half an integer
b) an integer
c) half an odd integer
d) an even integer plus 1/2

Answer: b Difficulty: 2

TRUE/FALSE

66. The intensity of a sound wave is proportional to its decibel level.

Answer: F Difficulty: 1

MULTIPLE CHOICE

67. The change in the frequency of a wave due to the motion of the source (or observer) is known as:
a) Mach effect
b) Bell effect
c) Doppler effect
d) interference

Answer: c Difficulty: 2

68. The sound source is moving away at Mach 1/2. What is the minimum speed with which the observer must move in the opposite direction to not receive sound waves?
a) Mach 0
b) Mach 0.5
c) Mach 1
d) Mach 2

Answer: c Difficulty: 2

Chapter 14

69. At low intensity levels, our perception of low and high sound frequencies (compared to mid frequencies) :
   a) is increased
   b) remains the same
   c) is decreased
   d) none of the above

   Answer: c Difficulty: 2

FILL-IN-THE-BLANK

70. What man made object exceeded the speed of sound long before the invention of the airplane or the bullet?

   Answer: A "cracked whip" creates a sonic boom. Difficulty: 2

MULTIPLE CHOICE

71. Middle C has a frequency of 262.Hz. What is the frquency of C an octave higher?
   a) 270.Hz
   b) 393.Hz
   c) 524.Hz
   d) 1310.Hz
   e) 2096.Hz
   f) 2620.Hz

   Answer: c Difficulty: 2

72. Two violin strings vibrating at 880.Hz and 876.Hz respectively, will cause a beat frequency of:
   a) 4.Hz
   b) 8.Hz
   c) 878.Hz
   d) 1756.Hz

   Answer: a Difficulty: 2

Chapter 15

MATCHING

Match the unit to the physical quantity.

a) dimensionless
b) volt
c) N-m²/C²
d) C²/N-m²
e) coulomb
f) N/C
g) farad
h) electron-volt

1. dielectric constant

    Answer: a) dimensionless   Difficulty: 2

2. electric potential

    Answer: b) volt   Difficulty: 2

3. Coulomb constant

    Answer: c) N-m²/C²   Difficulty: 2

4. permittivity

    Answer: d) C²/N-m²   Difficulty: 2

5. charge

    Answer: e) coulomb   Difficulty: 2

6. electric field

    Answer: f) N/C   Difficulty: 2

7. capacitance

    Answer: g) farad   Difficulty: 2

8. electric potential energy

    Answer: h) electron-volt   Difficulty: 2

Chapter 15

MULTIPLE CHOICE

9. Electrons carry a
   a) positive charge.
   b) negative charge.
   c) neutral charge.
   d) variable charge.

   Answer: b Difficulty: 2

10. Charge is
    a) quantized.
    b) conserved.
    c) invariant.
    d) all of the above.

    Answer: d Difficulty: 2

TRUE/FALSE

11. Quarks have greater charges than electrons.

    Answer: F Difficulty: 2

MULTIPLE CHOICE

12. The electron was discovered by
    a) Lord Rutherford.
    b) J.J. Thomson.
    c) Albert Einstein.
    d) Richard Feynman.

    Answer: b Difficulty: 2

13. The ratio of the neutron mass to the electron mass is approximately
    a) 2:1
    b) 20:1
    c) 200:1
    d) 2000:1

    Answer: d Difficulty: 2

14. A neutral atom always has
    a) more neutrons than protons.
    b) more protons than electrons.
    c) the same number of neutrons as protons.
    d) the same number of protons as electrons.

    Answer: d Difficulty: 2

15. A glass rod is rubbed with a piece of silk. During the process the glass rod acquires a positive charge and the silk
    a) acquires a positive charge also.
    b) acquires a negative charge.
    c) remains neutral.
    d) could either be positively charged or negatively charged. It depends on how hard the rod was rubbed.

    Answer: b Difficulty: 2

ESSAY

16. Consider an equilateral triangle of side 20 cm. A charge of +2μC is placed at one vertex and charges of -4μC are placed at the other two vertices. Determine the magnitude and direction of the electric field at the center of the triangle.

    Answer: 2.7×10^6 V/m Difficulty: 2

17. A metal sphere of radius 2 cm carries a charge of 3μC. What is the electric field 6 cm from the center of the sphere?

    Answer: 7.5×10^6 V/m Difficulty: 2

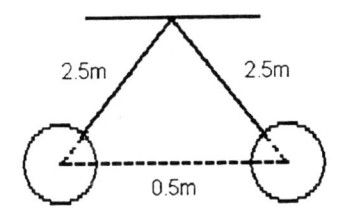

18. Two volleyballs, each of mass 0.3 kg, are charged by an electrostatic generator. Each is attached to an identical string and suspended from the same point. They repel each other and hang with separation 0.5 m. The length of the string from the point of support to the center of a ball is 2.5 m. Determine the charge on each ball.

    Answer: 2.86μC Difficulty: 2

19. Charge +2q is placed at the origin and charge -q is placed at x = 2a. Where can a third positive charge +q be placed so that the force on it is zero?

    Answer: x = 6.83 a Difficulty: 2

Chapter 15

MULTIPLE CHOICE

20. An electron and a proton are separated by a distance of 1 m. What happens to the size of the force on the proton if the electron is moved 0.5 m closer to the proton?
    a) It increases to 4 times its original value.
    b) It increases to 2 times its original value.
    c) It decreases to one-half its original value.
    d) It decreases to one-fourth its original value.

    Answer: a Difficulty: 2

21. The charge carried by one electron is $e = -1.6 \times 10^{-19}$ C. The number of excess electrons necessary to produce a charge of 1.0 C is
    a) 6.25×10^{18}
    b) 6.25×10^{9}
    c) 1.6×10^{19}
    d) none of the above.

    Answer: a Difficulty: 2

22. Sphere A carries a net charge and sphere B is neutral. They are placed near each other on an insulated table. Which statement best describes the electrostatic force between them?
    a) There is no force between them since one is neutral.
    b) There is a force of repulsion between them.
    c) There is a force of attraction between them.
    d) The force is attractive if A is charged positively and repulsive if A is charged negatively.

    Answer: c Difficulty: 2

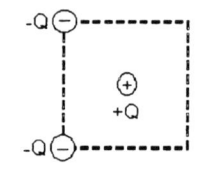

23. A point charge of +Q is placed at the center of a square, and a second point charge of -Q is placed at the upper-left corner. It is observed that an electrostatic force of 2 N acts on the positive charge at the center. What is the magnitude of the force that acts on the center charge if a third charge of -Q is placed at the lower-left corner?
    a) Zero
    b) $2\sqrt{2}$ N
    c) 4 N
    d) None of the above

    Answer: c Difficulty: 2

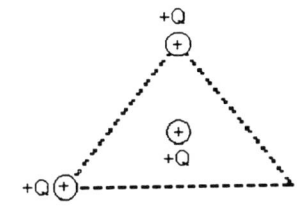

24. A point charge of +Q is placed at the centroid of an equilateral triangle. When a second charge of +Q is placed at one of the triangle's vertices, an electrostatic force of 4 N acts on it. What is the magnitude of the force that acts on the center charge when a third charge of +Q is placed at one of the other vertices?
   a) Zero
   b) 4 N
   c) 8 N
   d) None of the above

   Answer: b Difficulty: 2

25. Two charged objects attract each other with a certain force. If the charges on both objects are doubled with no change in separation, the force between them
   a) quadruples.
   b) doubles.
   c) halves.
   d) increases, but we can't say how much without knowing the distance between them.

   Answer: a Difficulty: 2

26. Two charged objects attract each other with a force F. What happens to the force between them if one charge is doubled, the other charge is tripled, and the separation distance between their centers is reduced to one-fourth its original value? The force is now equal to
   a) 16 F
   b) 24 F
   c) 6/16 F
   d) 96 F

   Answer: d Difficulty: 2

27. Consider point charges of +Q and +4Q, which are separated by 3 m. At what point, on a line between the two charges, would it be possible to place a charge of -Q such that the electrostatic force acting on it would be zero?
   a) There is no such point possible.
   b) 1 m from the +Q charge
   c) 1 m from the +4 charge
   d) 3/5 m from the +Q charge

   Answer: b Difficulty: 2

Chapter 15

28. Which of the following is an accurate statement?
    a) All parts of a perfect conductor are at the same potential.
    b) If a solid metal sphere carries a net charge, the charge will be uniformly distributed throughout the volume of the sphere.
    c) A conductor cannot carry a net charge.
    d) The electric field at the surface of a conductor is not necessarily perpendicular to the surface in all cases.

    Answer: a Difficulty: 2

TRUE/FALSE

29. The electric field at the surface of a conductor is always zero.

    Answer: F Difficulty: 2

MULTIPLE CHOICE

30. The concept of electric field lines was invented by
    a) James Clerk Maxwell.
    b) your Professor.
    c) Allesandro Volta.
    d) Michael Faraday.
    e) Isaac Newton.

    Answer: d Difficulty: 2

31. Electric field lines
    a) circle clockwise around positive charges.
    b) circle counter-clockwise around positive charges.
    c) radiate outward from negative charges.
    d) radiate outward from positive charges.

    Answer: d Difficulty: 2

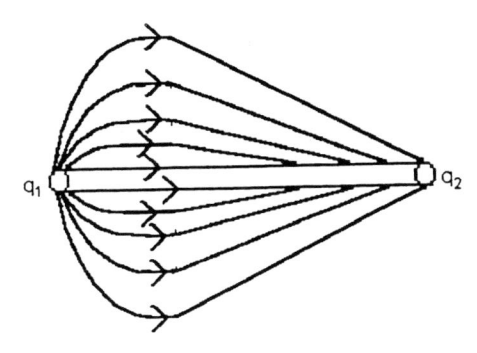

32. Two stationary point charges q_1 and q_2 are shown in the sketch along with some electric field lines representing the field between them. What can you deduce from the sketch?
   a) q_1 and q_2 have the same sign; the magnitudes are equal.
   b) q_1 and q_2 have the same sign; the magnitude of q_1 is greater than the magnitude of q_2.
   c) q_1 is positive and q_2 is negative; the magnitude of q_1 is greater than the magnitude of q_2.
   d) q_1 is negative and q_2 is positive; the magnitudes are equal.

   Answer: c Difficulty: 2

33. The electric field shown
   a) increases to the right.
   b) increases down.
   c) decreases to the right.
   d) decreases down.
   e) is uniform.

   Answer: a Difficulty: 2

34. A force of 6 N acts on a charge of 3μC when it is placed in a uniform electric field. What is the magnitude of this electric field?
   a) 18 μ-volts/m
   b) 2 μ-volts/m
   c) 0.5 μ-volts/m
   d) None of the above

   Answer: b Difficulty: 2

Chapter 15

35. A solid block of metal is placed in a uniform electric field. Which statement is correct concerning the electric field in the block's interior?
   a) The interior field points in a direction opposite to the exterior field.
   b) The interior field points in a direction that is at right angles to the exterior field.
   c) The interior points in a direction that is parallel to the exterior field.
   d) There is no electric field in the block's interior.

   Answer: d Difficulty: 2

36. If a solid metal sphere and a hollow metal sphere of equal diameters are each given the same charge, the electric field (E) midway between the center and the surface is
   a) greater for the solid sphere than for the hollow sphere.
   b) greater for the hollow sphere than for the solid sphere.
   c) zero for both.
   d) equal in magnitude for both, but one is opposite in direction from the other.

   Answer: c Difficulty: 2

37. A hollow metallic sphere is placed in a region permeated by a uniform electric field that is directed upward. Which statement is correct concerning the electric field in the sphere's interior?
   a) The field is zero everywhere in the interior.
   b) The field is directed upward.
   c) The field is directed downward.
   d) The field is zero only at the sphere's exact center.

   Answer: a Difficulty: 2

38. A positive charge is enclosed in a hollow metallic sphere that is not grounded. At a point directly above the hollow sphere, the electric field caused by the enclosed positive charge has
   a) diminished to zero.
   b) diminished somewhat.
   c) increased somewhat.
   d) not changed.

   Answer: d Difficulty: 2

39. A positive point charge is enclosed in a hollow metallic sphere that is grounded. At a point directly above the hollow sphere, the electric field caused by the enclosed positive charge has
   a) diminished to zero.
   b) diminished somewhat.
   c) increased somewhat.
   d) not changed.

   Answer: a Difficulty: 2

ESSAY

40. How much energy is necessary to place three charges, each of 2μC, at the corners of an equilateral triangle of side 2 cm?

    Answer: 5.4 J Difficulty: 2

TRUE/FALSE

41. Equipotential lines always begin and end on charges.

    Answer: F Difficulty: 2

MULTIPLE CHOICE

42. The lightning rod was invented by
    a) Dr. Frankenstein.
    b) Albert Einstein.
    c) Isaac Newton.
    d) Ben Franklin.
    e) Radio Shack.

    Answer: d Difficulty: 2

TRUE/FALSE

43. You cannot get an electric shock by touching a charged insulator.

    Answer: F Difficulty: 2

MULTIPLE CHOICE

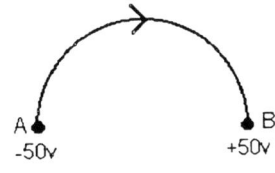

44. The net work done in moving an electron from point A at -50 V to point B at +50 V along the semi-circular path shown is
    a) $+1.6 \times 10^{-19}$ J
    b) -1.6×10^{-19} J
    c) zero
    d) cannot be determined; not enough information given.

    Answer: b Difficulty: 2

Chapter 15

ESSAY

45. Starting from rest, a proton falls through a potential difference of 1200 V. What speed does it acquire?

    Answer: 4.8×10^5 m/s Difficulty: 2

46. Four charges of equal charge +q are placed at the corners of a rectangle of sides a and b. What is the potential at the center of the rectangle if q = 2 μC, a = 3 cm, and b = 4 cm?

    Answer: 7.2×10^5 V Difficulty: 2

47. A metal sphere of radius 8 cm is charged to a potential of -500 V. With what velocity must an electron be fired toward the sphere if it is to just barely reach the sphere when started from a position 15 cm from the center of the sphere?

    Answer: 9×10^6 m/s Difficulty: 2

MULTIPLE CHOICE

48. It takes 10 J of energy to move 2 C of charge from point A to point B. What is the potential difference between points A and B?
    a) 20 V
    b) 0.2 V
    c) 5 V
    d) None of the above

    Answer: c Difficulty: 2

49. If a Cu^{2+} ion drops through a potential difference of 12 V, it will acquire a kinetic energy (in the absence of friction) of
    a) 12 eV
    b) 6 eV
    c) 24 eV
    d) none of the above.

    Answer: c Difficulty: 2

50. The electron-volt is a unit of
    a) voltage.
    b) current.
    c) power.
    d) energy.

    Answer: d Difficulty: 2

51. Consider a uniform electric field of 50 N/C directed toward the east. If the voltage measured relative to ground at a given point in the field is 80 V, what is the voltage at a point 1 m directly east of the point?
   a) 15 V
   b) 30 V
   c) 130 V
   d) Impossible to calculate from the information given.

   Answer: b Difficulty: 2

52. Consider a uniform electric field of 50 N/C directed toward the east. If the voltage measured relative to ground at a given point in the field is 80 V, what is the voltage at a point 1 m directly south of that point?
   a) zero
   b) 30 V
   c) 50 V
   d) 80 V

   Answer: d Difficulty: 2

53. The absolute potential at a distance of 2 m from a positive point charge is 100 V. What is the absolute potential 4 m away from the same point charge?
   a) 25 V
   b) 50 V
   c) 200 V
   d) 400 V

   Answer: b Difficulty: 2

54. The absolute potential at a distance of 2 m from a negative point charge is -100 V. What is the absolute potential 4 m away from the same point charge?
   a) -25 V
   b) -50 V
   c) -200 V
   d) -400 V

   Answer: b Difficulty: 2

55. The absolute potential at the center of a square is 3 V when a charge of +Q is located at one of the square's corners. What is the absolute potential at the square's center when a second charge of -Q is placed at one of the remaining corners?
   a) Zero
   b) 3 V
   c) 6 V
   d) None of the above

   Answer: a Difficulty: 2

Chapter 15

56. Which of the following is a vector?
   a) Electric potential
   b) Electric charge
   c) Electric field
   d) All of the above

   Answer: c Difficulty: 2

57. Several electrons are placed on a hollow sphere. They
   a) clump together on the sphere's outer surface.
   b) clump together on the sphere's inner surface.
   c) become uniformly distributed on the sphere's outer surface.
   d) become uniformly distributed on the sphere's inner surface.

   Answer: c Difficulty: 2

ESSAY

58. What charge appears on the plates of a 2 μF capacitor when it is charged to 100 V?

   Answer: 200 μC Difficulty: 2

59. A parallel plate capacitor is constructed with plate area of 0.4 m^2 and a plate separation of 0.1 mm. (a) How much charge is stored on it when it is charged to a potential difference of 12 V? (b) How much energy is stored?

   Answer: (a) 0.42 μC   (b) 30.6 μJ Difficulty: 2

MULTIPLE CHOICE

60. Electric Dipoles always consist of two charges that are
   a) equal in magnitude; opposite in sign.
   b) equal in magnitude; both are negative.
   c) equal in magnitude; both are positive.
   d) unequal in magnitude; opposite in sign.

   Answer: a Difficulty: 2

ESSAY

61. Capacitances of 10 μF and 20 μF are connected in parallel, and this pair is then connected in series with a 30 μF capacitor. What is the equivalent capacitance of this arrangement?

   Answer: 15 μF Difficulty: 2

62. Consider capacitors C_1, C_2, and C_3, which are connected in series in a closed loop. A switch is placed between C_1 and C_2. With the switch open, C_1 is charged to 12 volts by a battery. The battery is then disconnected and the switch is closed. Determine the final charge on each capacitor, and the potential difference across each given that $C_1 = 2$ μF, and $C_2 = C_3 = 3$ μF.

Answer:
$Q_1 = 13.7$ μF; $V_1 = 6.86$ V
$Q_2 = Q_3 = 10.3$ μC; $V_2 = V_3 = 3.43$ V
Difficulty: 2

MULTIPLE CHOICE

63. A battery charges a parallel-plate capacitor fully and then is removed. The plates are immediately pulled apart. (With the battery disconnected, the amount of charge on the plates remains constant.) What happens to the potential difference between the plates as they are being separated?
a) It increases.
b) It decreases.
c) It remains constant.
d) There is no way to tell from the information given.

Answer: a Difficulty: 2

64. The plates of a parallel-plate capacitor are maintained with constant voltage by a battery as they are pulled apart. During this process, the amount of charge on the plates must
a) increase.
b) decrease.
c) remain constant.
d) either increase or decrease. There is no way to tell from the information given.

Answer: b Difficulty: 2

65. The plates of a parallel-plate capacitor are maintained with constant voltage by a battery as they are pulled apart. What happens to the strength of the electric field during this process?
a) It increases.
b) It decreases.
c) It remains constant.
d) There is no way to tell from the information given.

Answer: b Difficulty: 2

Chapter 15

66. A parallel-plate capacitor has plates of area 0.2 m² separated by a distance of 0.001 m. What is the strength of the electric field between these plates when this capacitor is connected to a 6-V battery?
   a) 1200 N/C
   b) 3000 N/C
   c) 6000 N/C
   d) None of the above

   Answer: c   Difficulty: 2

67. A parallel-plate capacitor has plates of area 0.2 m² separated by a distance of 0.001 m. What is this capacitor's capacitance?
   a) 200 F
   b) 40 F
   c) 1.77×10^{-9} F
   d) 3.54×10^{-10} F

   Answer: c   Difficulty: 2

68. A charge of 2 µC flows onto the plates of a capacitor when it is connected to a 12-V battery. How much work was done in charging this capacitor?
   a) 24 µJ
   b) 12 µJ
   c) 144 µJ
   d) None of the above

   Answer: b   Difficulty: 2

69. A 15-µF capacitor is connected to a 50-V battery and becomes fully charged. The battery is removed and a slab of dielectric that completely fills the space between the plates is inserted. If the dielectric has a dielectric constant of 5, what is the capacitance of the capacitor after the slab is inserted?
   a) 75 µF
   b) 20 µF
   c) 3 µF
   d) None of the above

   Answer: a   Difficulty: 2

70. A 15-µF capacitor is connected to a 50-V battery and becomes fully charged. The battery is removed and a slab of dielectric that completely fills the space between the plates is inserted. If the dielectric has a dielectric constant of 5, what is the voltage across the capacitor's plates after the slab is inserted?
   a) 250 V
   b) 10 V
   c) 2 V
   d) None of the above

   Answer: b   Difficulty: 2

71. A neutral atom has
    a) equal numbers of electrons and protons
    b) equal numbers of electrons and neutrons
    c) equal numbers of protons and neutrons
    d) equal numbers of electrons, protons, and neutrons

    Answer: a Difficulty: 1

72. Which of the following does not have dimensions of an electric field?
    a) N/C
    b) Kg-m/(C-s^2)
    c) volts/meter
    d) Farads/C

    Answer: d Difficulty: 1

73. An electron which moves from the negative to the positive terminal of a 2.volt battery loses how much potential energy?
    a) 2.J
    b) 2x10^{-19} J
    c) 3.2x10^{-19} J
    d) 2.eV
    e) none of the above

    Answer: d Difficulty: 2

FILL-IN-THE-BLANK

74. How far apart should two protons be if the electrical force of repulsion is equal to the weight (attraction to the Earth) of one of the protons?

    Answer: 12.cm Difficulty: 2

ESSAY

75. A charged rod will attract a thin stream of water falling from a faucet even though the water is neutral, it has no net charge. How can the rod attract the water?

    Answer:
    The Coulomb Force of the rod polarizes each molecule of water (one sign of charge is attracted and the opposite sign is repelled). The charge that is attracted feels a slightly larger attraction than the opposite charge is repelled because it is at a slightly different distance from the rod.
    Difficulty: 2

Chapter 15

76. What does it mean to say something (for example charge) is "quantized"?

    Answer:
    A quantized something comes in basic units and occurs as an integer number of these basic units. For example, isolated objects will be found to have a charge plus or minus ne where n is an integer and $e=1.6\times10^{-19}$ Coulombs
    Difficulty: 1

MULTIPLE CHOICE

77. The dielectric constant of a vacuum is
    a) E_0
    b) 8.99×10^9
    c) 0.000
    d) 1.000

    Answer: d Difficulty: 1

78. In electricity, what quantity is analogous to the "acceleration of gravity" g (which is a force per unit mass):
    a) electric force
    b) electric field
    c) electric potential
    d) electric charge

    Answer: b Difficulty: 2

79. The force which binds, or holds, atoms together to form molecules is
    a) magnetic
    b) electrical
    c) gravitational
    d) nuclear
    e) friction

    Answer: b Difficulty: 1

ESSAY

80. Only gravitation determines the planetary orbits. Why aren't the other fundamental forces (and electromagnetism in particular) important for the motions?

    Answer:
    Only gravity and electromagnetism are long range, however the planets and satellites are electrically neutral. The planets do not have "negative mass" to neutralize the positive mass so gravity predominates.
    Difficulty: 2

MULTIPLE CHOICE

81. When an electron is removed from a neutral atom, it becomes:
    a) a positive ion
    b) a negative ion
    c) a bipolar atom
    d) heavier
    e) none of the above

    Answer: a   Difficulty: 2

TRUE/FALSE

82. The electric constant (k in kqQ/r^2) is negative for negative charge and positive for positive charge.

    Answer: F   Difficulty: 1

83. Like charges repel, unlike charges attract.

    Answer: T   Difficulty: 1

84. A mole contains Avogadro's number of protons.

    Answer: F   Difficulty: 1

85. Since there is a universal Law of Conservation of Charge, charges can not be created or destroyed.

    Answer: F   Difficulty: 1

ESSAY

86. Explain how objects can get charged by FRICTION.

    Answer:
    When two dissimilar materials are rubbed together, one literally rubs or tears electrons off the one type of molecule, leaving it positively charged, and the other material acquires the extra electrons, thereby becoming negative.

    Difficulty: 2

FILL-IN-THE-BLANK

87. There is a 5.0 µC charge at each of 3 corners of a square (each side 70.mm long). What is the force on +6.0 µC placed at the center of the square?

    Answer: 28.Newtons toward the empty corner   Difficulty: 2

Chapter 15

ESSAY

88. Ben Franklin named charges "plus" and "minus". Suppose he had gotten it reversed and had a called plus charges "minus" and called minus "plus", what would have been the consequences?

    Answer:
    Like charges would still repel, unlike charges would attract. Coulomb's Law would still describe the electric force.

    Difficulty: 2

FILL-IN-THE-BLANK

89. Consider a container of 2.0 grams of hydrogen (one gram mole). Suppose you removed all the electrons and moved then to the other side of the Earth (diameter 12740.Km).
    A) How much charge is left behind?
    B) What is the attractive force between the protons here and the electrons at the other side of the Earth?

    Answer: A) 1.9×10^5 Coulombs   B) 2.1×10^6 Newtons = 2.1×10^2 tons!

    Difficulty: 2

90. Consider the x-axis with 4.0 Coulomb at x=1.0 meter and -2.0 C at the origin. Where should 6.0 C be placed so that the net force on it is zero?

    Answer: x = -2.4 meter   Difficulty: 2

Chapter 16

MATCHING

Match the unit to the physical quantity.

a) ohm
b) $(\Omega\text{-m})^{-1}$
c) $(°C)^{-1}$
d) ampere
e) $\Omega\text{-m}$
f) volt-amp

1. electric resistance

    Answer: a) ohm Difficulty: 2

2. conductivity

    Answer: b) $(\Omega\text{-m})^{-1}$ Difficulty: 2

3. temperature coefficient of resistivity

    Answer: c) $(°C)^{-1}$ Difficulty: 2

4. electric current

    Answer: d) ampere Difficulty: 2

5. resistivity

    Answer: e) $\Omega\text{-m}$ Difficulty: 2

6. electric power

    Answer: f) volt-amp Difficulty: 2

MULTIPLE CHOICE

7. A car battery
    a) has an emf of 6 V consisting of one 6-V cell.
    b) has an emf of 6 V consisting of three 2-V cells connected in series.
    c) has an emf of 6 V consisting of three 2-V cells connected in parallel.
    d) has an emf of 12 V consisting of six 2-V cells connected in series.
    e) has an emf of 12 V consisting of six 2-V cells connected in parallel.

    Answer: d Difficulty: 2

Chapter 16

8. If you connect two identical storage batteries together in parallel, and place them in a circuit, the combination will provide
    a) twice the voltage and twice the total charge that one battery would.
    b) twice the voltage and the same total charge that one battery would.
    c) the same voltage and twice the total charge that one battery would.
    d) half the voltage and half the total charge that one battery would.
    e) half the voltage and twice the total charge that one battery would.

    Answer: c Difficulty: 2

9. If you connect two identical storage batteries together in series ("+" to "-"}" to "+" to "-"), and place them in a circuit, the combination will provide
    a) zero volts.
    b) twice the voltage, and different currents will flow through each.
    c) twice the voltage, and the same current will flow through each.
    d) the same voltage, and different currents will flow through each.
    e) the same voltage and the same current will flow through each.

    Answer: c Difficulty: 2

10. Car batteries are rated in "amp-hours." This is a measure of their
    a) charge.
    b) current.
    c) emf.
    d) power.

    Answer: a Difficulty: 2

TRUE/FALSE

11. The net direction in which electrons flow through a circuit is called conventional current.

    Answer: F Difficulty: 2

ESSAY

12. A charge of 12 C passes through an electroplating apparatus in 2 min. What is the average current?

    Answer: 0.1 A Difficulty: 2

13. A battery is rated at 12 V and 160 A-h. How much charge does the battery store?

    Answer: 48,000 J Difficulty: 2

14. The diameter of no. 12 copper wire is 0.081 in. The maximum safe current it can carry (in order to prevent fire danger in building construction) is 20 A. At this current, what is the drift velocity of the electrons? (The number of electron carriers in one cubic centimeter of copper is 8.5×10^{22}.)

    Answer: 0.44 mm/s Difficulty: 2

15. In an electroplating process, it is desired to deposit 40 mg of silver on a metal part by using a current of 2 A. How long must the current be allowed to run to deposit this much silver? (The silver ions are singly charged, and the atomic weight of silver is 108.)

    Answer: 17.85 s Difficulty: 2

MULTIPLE CHOICE

16. A coulomb per second is the same as
    a) a watt.
    b) an ampere.
    c) a volt-second.
    d) a volt per second.

    Answer: b Difficulty: 2

17. Consider two copper wires each carrying a current of 3 A. One wire has twice the diameter of the other. The ratio of the drift velocity in the smaller diameter wire to that in the larger diameter wire is
    a) 4:1
    b) 2:1
    c) 1:2
    d) 1:4

    Answer: b Difficulty: 2

ESSAY

18. A coffee maker, which draws 13.5 A of current, has been left on for 10 min. What is the net number of electrons that have passed through the coffee maker?

    Answer: 5.06×10^{22} Difficulty: 2

TRUE/FALSE

19. A device obeying Ohm's law must have a constant resistance.

    Answer: T Difficulty: 2

Chapter 16

20. A potential difference (V) is applied to the ends of an object causing a current (I) to flow through it. The object obeys Ohm's law because we can obtain the resistance by dividing the voltage by the current: R = V/I.

    Answer: F Difficulty: 2

21. A lightbulb is an example of a device that obeys Ohm's law.

    Answer: F Difficulty: 2

ESSAY

22. What potential difference is required to cause 2 A to flow through a resistance of 8 Ω?

    Answer: 16 V Difficulty: 2

23. What is the resistance of 1.0 m of no. 18 copper wire (diameter 0.40")? (The resistivity of copper is 1.69×10^{-8} Ω-m.)

    Answer: 0.00065 Ω Difficulty: 2

24. A heavy bus bar is 20 cm long and of rectangular cross-section, 1 cm x 2 cm. What is the voltage drop along its length when it carries 4000 A? (The resistivity of copper is 1.69×10^{-8} Ω-m.)

    Answer: 0.068 V Difficulty: 2

25. The temperature coefficient of resistivity of platinum is $3.9 \times 10^{-3}/°K$. If a platinum wire has a resistance of R at room temperature (23° C), to what temperature must it be heated in order to double its resistance to 2R?

    Answer: 279° C Difficulty: 2

MULTIPLE CHOICE

26. The resistance of a wire is defined as
    a) (length)(resistivity)/(cross-sectional area).
    b) (current)/(voltage).
    c) (voltage)/(current).
    d) none of the above.

    Answer: c Difficulty: 2

27. Consider two copper wires. One has twice the length and twice the cross-sectional area of the other. How do the resistances of these two wires compare?
    a) Both wires have the same resistance.
    b) The longer wire has twice the resistance of the shorter wire.
    c) The longer wire has four times the resistance of the shorter wire.
    d) None of the above.

    Answer: a Difficulty: 2

28. Consider two copper wires. One has twice the length of the other. How do the resistivities of these two wires compare?
    a) Both wires have the same resistivity.
    b) The longer wire has twice the resistivity of the shorter wire.
    c) The longer wire has four times the resistivity of the shorter wire.
    d) None of the above.

    Answer: a Difficulty: 2

29. Negative temperature coefficients of resistivity
    a) do not exist.
    b) exist in conductors.
    c) exist in semiconductors.
    d) exist in superconductors.

    Answer: c Difficulty: 2

30. Which of the following graphs indicates the behavior of a superconductor?

    Answer: c Difficulty: 2

Chapter 16

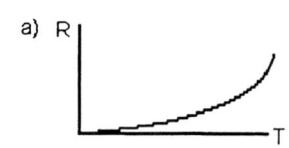

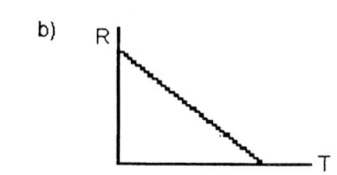

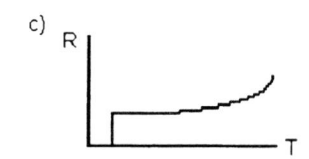

a) graph showing R vs T, R increasing exponentially
b) graph showing R vs T, R decreasing linearly
c) graph showing R vs T, R stepped then increasing
d) graph showing R vs T, R constant

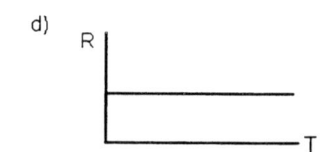

31. Which of the following graphs indicates that the material is a semiconductor?

    Answer: b Difficulty: 2

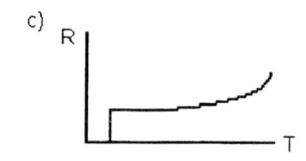

a) graph showing R vs T, R increasing exponentially
b) graph showing R vs T, R decreasing linearly
c) graph showing R vs T, R stepped then increasing
d) graph showing R vs T, R constant

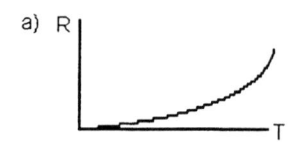

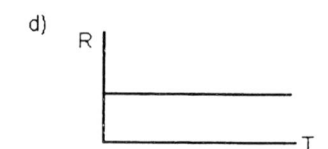

32. Which of the following graphs indicates that Ohm's law is obeyed for the range shown?

    Answer: d Difficulty: 2

33. If the resistance in a constant voltage circuit is doubled, the power dissipated by that circuit will
    a) increase by a factor of two.
    b) increase by a factor of four.
    c) decrease to one-half its original value.
    d) decrease to one-fourth its original value.

    Answer: c Difficulty: 2

34. The resistivity of most common metals
   a) remains constant over wide temperature ranges.
   b) increases as the temperature increases.
   c) decreases as the temperature increases.
   d) varies randomly as the temperature increases.

   Answer: b Difficulty: 2

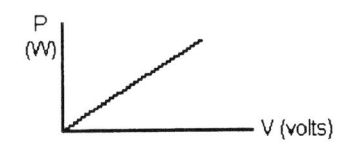

35. In the graph shown, what physical quantity does the slope represent?
   a) Current
   b) Energy
   c) Resistivity
   d) Resistance

   Answer: a Difficulty: 2

36. A 500-W device is connected to a 100-V power source. What current flows through this surface?
   a) 50,000 A
   b) 0.2 A
   c) 5 A
   d) None of the above

   Answer: c Difficulty: 2

ESSAY

37. A lamp uses a 150-W bulb. If it is used at 120 V, what current does it draw and what is its resistance?

   Answer: 1.25 A; 96 Ω Difficulty: 2

38. A 200-Ω resistor is rated at 1/4 W. (a) What is the maximum current it can draw? (b) What is the maximum voltage that should be applied across it?

   Answer: (a) 0.0354 A   (b) 7.07 V Difficulty: 2

39. A motor that can do work at the rate of 2 hp has 60% efficiency. How much current does it draw from a 120 V line? (1 hp = 746 W.)

   Answer: 20.7 A Difficulty: 2

Chapter 16

40. The monthly (30 days) electric bill included the cost of running a central air-conditioning unit for 2 hr/day at 5000 W, and a series connection of ten 60-W lightbulbs for 5 hr/day. How much did these items contribute to the cost of the monthly electric bill if electricity costs 8τ per kWh?

    Answer: $31.20 Difficulty: 2

41. The heating element in an electric drier operates on 240 V and generates heat at the rate of 2 kW. The heating element shorts out and, in repairing it, the owner shortens the Nichrome wire by 10%. (Assume the temperature is unchanged. In reality, the resistivity of the wire will depend on its temperature.) What effect will the repair have on the power dissipated in the heating element?

    Answer: Power dissipated increases to 2.22 kW Difficulty: 2

MULTIPLE CHOICE

42. A battery has an internal resistance of 2 Ω. This battery delivers maximum power to a load resistor that has a value of
    a) zero
    b) 1 Ω
    c) 2 Ω
    d) 4 Ω

    Answer: c Difficulty: 2

43. Appliances in the USA are designed to work on 110-120 V, whereas appliances in Europe operate on 220-240 V. What would happen if you tried to use a European electric shaver in this country?
    a) It would probably work as well as it does in Europe, since it is the power rating, not the voltage rating, that matters.
    b) It would probably work as well as it does in Europe, since it is the current rating, not the voltage rating, that matters.
    c) It would barely work (or maybe not work at all), since the voltage here is too low to push enough current through the device.
    d) It would probably overheat and burn up before very long.
    e) This question cannot be answered without knowing the frequency used in Europe and in the USA, since this is the parameter that matters most in any electrical device that operates on alternating current.

    Answer: c Difficulty: 2

44. A capacitor acquires and stores:
    a) charge
    b) potential
    c) energy
    d) all of the above

    Answer: d Difficulty: 2

FILL-IN-THE-BLANK

45. How might one connect four 10.µF capacitors to create one equivalent capacitor having a capacitance of 10.µF?

    Answer:
    Connect two in series and place that in parallel with two other in series.
    Difficulty: 2

TRUE/FALSE

46. Electric force per unit charge is called electric potential.

    Answer: F   Difficulty: 1

MULTIPLE CHOICE

47. A voltage has been applied across a capacitor. If the dielectric is replaced with another dielectric constant **eight** times as great and the voltage is reduced to **half** of what it was, the ENERGY STORED in the capacitor is how many times the original stored energy?
    a) 1/2
    b) 1/4
    c) 8
    d) 4
    e) 2
    f) 1

    Answer: e   Difficulty: 2

48. When a battery is connected to a capacitor, the capacitor charges to 99% of its final value in 0.05 s. What is the resistance in the circuit?
    a) 2.mΩ
    b) 2.Ω
    c) 2.KΩ
    d) 2.MΩ

    Answer: c   Difficulty: 2

Chapter 16

FILL-IN-THE-BLANK

49. A 4.0µF and a 1.0µF capacitor are connected in parallel and both are charged with a 12.volt battery. They are disconnected from the battery and reconnected plus side of the 1st to negative side of 2nd, and neg. side of 1st to pos. side of 2nd.
    a) how much was each charged when connected to the battery?
    b) after reconnecting, how much charge is stored ?
    c) What potential exists across the final parallel combination ?

    Answer: a) 48.µCoulomb & 12.µCoulomb , b) 36.µC , c) 7.1 volts

    Difficulty: 2

50. A parallel plate capacitor is constructed with Teflon (dielectric constant 2.1) between the plates. This 12.6 µF has been charged to 1.5 volts. The Teflon is then pulled out (removed).
    a) What charge was originally stored ?
    b) After removing the Teflon, what potential is across the plates?

    Answer: a) 6.0 µC   b) 3.2 volts   Difficulty: 2

ESSAY

51. Why is it dangerous to probe inside a television set even hours after it has been unplugged?

    Answer:
    Capacitors are charged to large potentials and store considerable charge within the TV sets and they still hold charge long after being disconnected from the power supply.

    Difficulty: 2

52. A charged capacitor stores energy and when the Teflon dielectric between the plates is pulled out the energy stored more than doubles. Where did the extra energy come from?

    Answer:
    The charged plates have a net attraction to the polarized charges in the dielectric as the dielectric is removed. We must do work to remove the Teflon which puts extra energy into the capacitor system.

    Difficulty: 2

FILL-IN-THE-BLANK

53. After being charged from a battery, the plates of a parallel plate capacitor are moved closer together. When they are half as far apart as originally, by how much does the stored energy change?

    Answer:
    The capacity doubles but charge remains the same so energy ($Q^2/2C$) reduces to half of what it was.
    Difficulty: 2

ESSAY

54. If you were a parallel plate capacitor manufacturer, state the various ways you might make larger valued capacitors.

    Answer:
    Increase plate area, decrease separation, and/or increase the dielectric constant.
    Difficulty: 2

55. If you find you are able to move a positive charge from infinity (far away) to a particular point without doing any work, can you infer that no other charges are nearby?

    Answer:
    NO. For example, consider 2 rows of opposite charge. If you had moved half way between the charges, no work would have been done.
    Difficulty: 2

TRUE/FALSE

56. A proton (being 1836 times heavier than an electron) gains 1836.eV when moving through a potential increase of one volt.

    Answer: T Difficulty: 1

57. The electron-volt is only used to describe energies of ELECTRONS.

    Answer: F Difficulty: 1

58. Capacitors connected in series always have less total capacitance than any of the individual capacities.

    Answer: T Difficulty: 1

Chapter 16

FILL-IN-THE-BLANK

59. An electron moving from the negative terminal to the positive terminal of a 12. volt battery gains (or loses) how much energy? Be sure to indicate whether it is a gain or loss.

    Answer: Loses 12.eV (1.9×10^{-19} J) Difficulty: 2

MULTIPLE CHOICE

60. An electron volt is a unit of:
    a) potential
    b) charge
    c) voltage
    d) energy

    Answer: d Difficulty: 1

61. An electric field has units of:
    a) V-m
    b) J/C
    c) J/m
    d) V/m
    e) N-m

    Answer: d Difficulty: 1

62. A negative charge, if free, tries to move
    a) from high potential to low potential
    b) from low potential to high potential
    c) toward infinity
    d) away from infinity

    Answer: b Difficulty: 1

ESSAY

63. Why can, or can not, electric field lines intersect?

    Answer:
    The electric field line is in the direction of the force on a positive "test charge" placed at the point. The force can not point in two directions at one point.

    Difficulty: 2

FILL-IN-THE-BLANK

64. Thirty microcoulombs of NEGATIVE charge experiences an electrostatic force of 27.mN. What is the magnetude and direction of the electric field?

Answer: 0.90 kN/C (or kV/m) in the direction opposite the force.

Difficulty: 2

Chapter 17

MATCHING

Match the device to the electrical symbol.

a) ground   d) capacitor   g) fuse
b) voltmeter   e) battery   h) switch
c) resistor   f) galvanometer   i) ammeter

1.

   Answer: a) ground   Difficulty: 2

2.

   Answer: b) voltmeter   Difficulty: 2

3.

   Answer: c) resistor   Difficulty: 2

4.

   Answer: d) capacitor   Difficulty: 2

5.

   Answer: e) battery   Difficulty: 2

6.

   Answer: f) galvanometer   Difficulty: 2

Testbank

7.

Answer: g) fuse   Difficulty: 2

8.

Answer: h) switch   Difficulty: 2

9.

Answer: i) ammeter   Difficulty: 2

MULTIPLE CHOICE

10. When two or more resistors are connected in series to a battery
    a) the total voltage across the combination is the algebraic sum of the voltages across the individual resistors.
    b) the same current flows through each resistor.
    c) the equivalent resistance of the combination is equal to the sum of the resistances of each resistor.
    d) all of the above.

    Answer: d   Difficulty: 2

11. When two or more resistors are connected in parallel to a battery,
    a) the voltage across each resistor is the same.
    b) the total current flowing from the battery equals the sum of the currents flowing through each resistor.
    c) the equivalent resistance of the combination is less than the resistance of any one of the resistors.
    d) all of the above.

    Answer: d   Difficulty: 2

12. Three identical resistors are connected in series to a battery. If the current of 12 A flows from the battery, how much current flows through any one of the resistors?
    a) 12 A
    b) 4 A
    c) 36 A
    d) None of the above

    Answer: a   Difficulty: 2

Chapter 17

13. Three identical resistors are connected in parallel to a battery. If the current of 12 A flows from the battery, how much current flows through any one of the resistors?
a) 12 A
b) 4 A
c) 36 A
d) None of the above

Answer: b Difficulty: 2

14. Three identical resistors are connected in series to a 12-V battery. What is the voltage across any one of the resistors?
a) 36 V
b) 12 V
c) 4 V
d) None of the above

Answer: c Difficulty: 2

15. Three identical resistors are connected in parallel to a 12-V battery. What is the voltage of any one of the resistors?
a) 36 V
b) 12 V
c) 4 V
d) None of the above

Answer: b Difficulty: 2

16. A 6-Ω and a 12-Ω resistor are connected in series to a 36-V battery. What power is dissipated by the 6-Ω resistor?
a) 216 W
b) 48 W
c) 486 W
d) None of the above

Answer: d Difficulty: 2

17. A 6-Ω and a 12-Ω resistor are connected in parallel to a 36-V battery. What power is dissipated by the 6-Ω resistor?
a) 216 W
b) 48 W
c) 486 W
d) None of the above

Answer: a Difficulty: 2

18. As more resistors are added in series to a constant voltage source, the power supplied by the source
   a) increases.
   b) decreases.
   c) does not change.
   d) increases for a time and then starts to decrease.

   Answer: b Difficulty: 2

19. As more resistors are added in parallel to a constant voltage source, the power supplied by the source
   a) increases.
   b) decreases.
   c) does not change.
   d) increases for a time and then starts to decrease.

   Answer: a Difficulty: 2

TRUE/FALSE

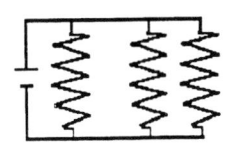

20. If identical resistors are connected in parallel, and a battery is connected as shown, the resistor closest to the battery will have the largest current flowing through it.

   Answer: F Difficulty: 2

MULTIPLE CHOICE

21. You obtain a 100-W lightbulb and a 50-W light bulb. Instead of connecting them in the normal way, you devise a circuit that places them in series across normal household voltage. Which statement is correct?
   a) Both bulbs glow at the same reduced brightness.
   b) Both bulbs glow at the same increased brightness.
   c) The 100-W bulb glows brighter than the 50-W bulb.
   d) The 50-W bulb glows more brightly than the 100-W bulb.

   Answer: d Difficulty: 2

Chapter 17

22. The lamps in a string of Christmas tree lights are connected in parallel. What happens if one lamp burns out? (Assume negligible resistance in the wires leading to the lamps.)
   a) The brightness of the lamps will not change appreciably.
   b) The other lamps get brighter equally.
   c) The other lamps get brighter, but some get brighter than others.
   d) The other lamps get dimmer equally.
   e) The other lamps get dimmer, but some get dimmer than others.

   Answer: e Difficulty: 2

ESSAY

23. Four 20-Ω resistors are connected in parallel. What is the equivalent resistance?

   Answer: 5 Ω Difficulty: 2

24. A combination of 2 Ω in series with 4 Ω is connected in parallel with 3 Ω. What is the equivalent resistance?

   Answer: 2 Ω Difficulty: 2

25. Two 100-W lightbulbs are to be connected to a 120 V source. What is the current, potential difference, and dissipated power for each when they are connected (a) in parallel (the normal arrangement)? (b) in series?

   Answer:
   (a) 0.83 A in each; 120 V; 100 W each, 20 W total.
   (b) 0.42 A in each; 60 V; 25 W

   Difficulty: 2

26. What different resistances can be obtained by using two 2-Ω resistors and one 4-Ω resistor?

   Answer: 0.8, 1.5, 2.0, 2.3, 5, 8 Ω Difficulty: 2

MULTIPLE CHOICE

27. Two 4-Ω resistors are connected in parallel, and this combination is connected in series with 3 Ω. What is the effective resistance of this combination?
   a) 1.2 Ω
   b) 5 Ω
   c) 7 Ω
   d) 11 Ω

   Answer: b Difficulty: 2

28. A 3-Ω resistor is connected in parallel with a 6-Ω resistor. This combination is connected in series with a 4-Ω resistor. The resistors are connected to a 12-volt battery. How much power is dissipated in the 3-Ω resistor?
   a) 2.7 W
   b) 5.3 W
   c) 6 W
   d) 12 W

   Answer: b Difficulty: 2

29. When resistors are connected in parallel, we can be certain that
   a) the same current flows in each one.
   b) the potential difference across each is the same.
   c) the power dissipated in each is the same.
   d) their equivalent resistance is greater than the resistance of any one of the individual resistances.

   Answer: b Difficulty: 2

30. When resistors are connected in series,
   a) the same power is dissipated in each one.
   b) the potential difference across each is the same.
   c) the current flowing in each is the same.
   d) more than one of the above is true.

   Answer: c Difficulty: 2

31. Consider three identical resistors, each of resistance R. The maximum power each can dissipate is P. Two of the resistors are connected in series, and a third is connected in parallel with these two. What is the maximum power this network can dissipate?
   a) 2P/3
   b) 3P/2
   c) 2P
   d) 3P

   Answer: b Difficulty: 2

32. Four 2-Ω resistors are connected to form a regular tetrahedron. What is the resistance between two vertices of the tetrahedron?
   a) 1 Ω
   b) 1.5 Ω
   c) 2 Ω
   d) 2.4 Ω

   Answer: a Difficulty: 2

Chapter 17

33. What is the maximum number of different resistance values obtainable by using three resistors?
    a) 6
    b) 9
    c) 11
    d) 17

    Answer: d Difficulty: 2

34. A 3-Ω resistor is connected in parallel with a 6-Ω resistor. This pair is then connected in series with a 4-Ω resistor. These resistors are connected to a battery. What will happen if the 3-Ω resistor breaks, i.e., becomes an infinite resistance?
    a) The current in the 4-Ω resistor will drop to zero.
    b) The current in the 6-Ω resistor will increase.
    c) The current provided by the battery will not change.
    d) The power dissipated in the circuit will increase.

    Answer: b Difficulty: 2

35. One terminal of a battery is connected to a closed circuit, while the other terminal is left unconnected. Which of the following statements is true?
    a) No current flows through the battery, and the potential of any point in the circuit is unchanged.
    b) No current flows through the battery, and the potential of any point in the circuit is different.
    c) Current flows through the battery, and the potential of any point in the circuit is unchanged.
    d) Current flows through the battery, and the potential of any point in the circuit is different.

    Answer: b Difficulty: 2

ESSAY

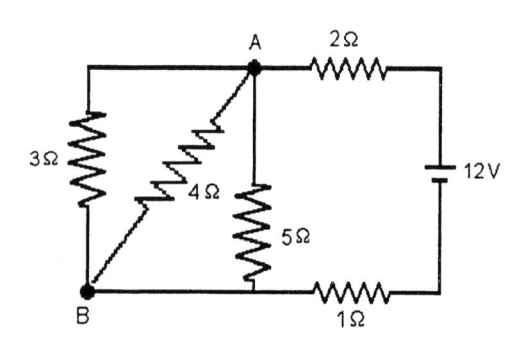

36. For the circuit shown, determine: (a) the current in each resistor b) the potential difference between points A and B.

    Answer:
    (a) 2.8 A in 1-Ω and 2-Ω resistors, 1.19 A in 3-Ω resistor, 0.90 A in 4-Ω resistor, 0.72 A in 5-Ω resistor
    (b) 3.58 V between A and B

    Difficulty: 2

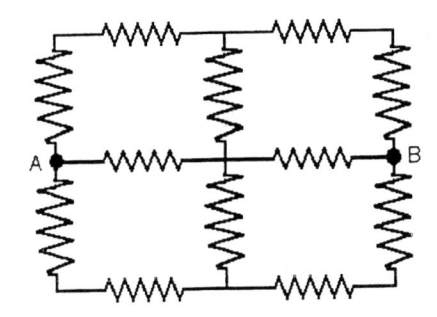

37. Determine the effective resistance between terminals A and B for the circuit shown here. Each resistor is 10 Ω.

    Answer: 10 Ω   Difficulty: 2

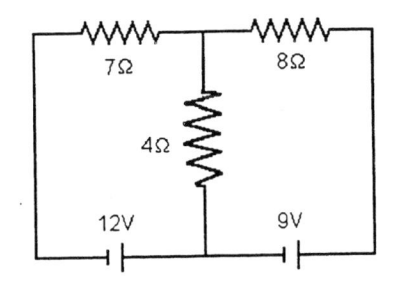

38. Determine the current and its direction, in each resistor, for the circuit shown here.

    Answer: 0.20 A to left in 7 Ω, 0.20 A to right in 8 Ω, 0.40 A up in 4 Ω

    Difficulty: 2

MULTIPLE CHOICE

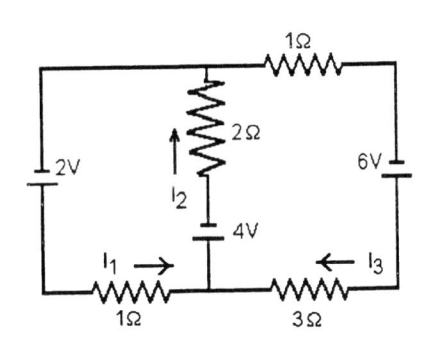

39. Which of the equations here is valid for the circuit shown?
    a) $2 - I_1 - 2I_2 = 0$
    b) $2 - 2I_1 - 2I_2 - 4I_3 = 0$
    c) $4 - I_1 + 4I_3 = 0$
    d) $-2 - I_1 - 2I_2 = 0$
    e) $6 - I_1 - 2I_2 = 0$

    Answer: c   Difficulty: 2

Chapter 17

40. Kirchhoff's junction rule is an example of
   a) conservation of energy.
   b) conservation of charge.
   c) conservation of momentum.
   d) none of the above.

   Answer: b Difficulty: 2

41. Kirchhoff's voltage rule for a closed loop is an example of
   a) conservation of energy.
   b) conservation of charge.
   c) conservation of momentum.
   d) none of the above.

   Answer: a Difficulty: 2

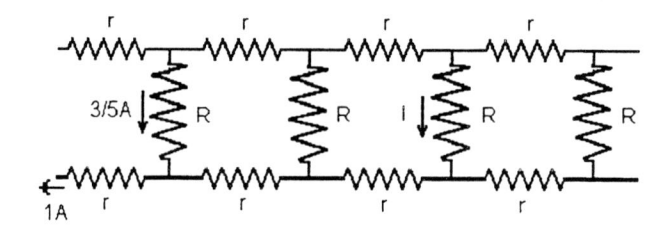

42. Shown here are a few segments of a circuit used to model an infinitely long transmission line. The circuit continues indefinitely in both directions. What is the value, in amperes, of the current I?
   a) 8/125
   b) 12/125
   c) 16/75
   d) 4/25
   e) 2/25

   Answer: b Difficulty: 2

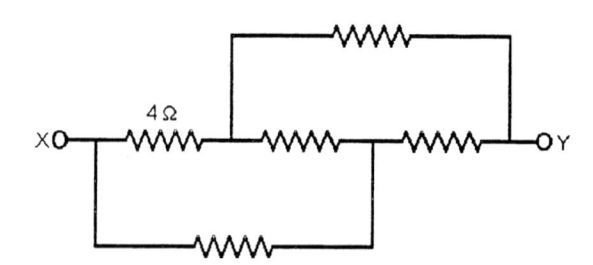

43. What is the resistance between terminals X and Y for the circuit shown here?
   a) 2 Ω
   b) 2.4 Ω
   c) 3.3 Ω
   d) 5.7 Ω

   Answer: c Difficulty: 2

ESSAY

44. A 2-μF capacitor is charged to 12 V and then discharged through 4×10^6 Ω. How long will it take for the voltage across the capacitor to drop to 3 V?

    Answer: 5.55 s Difficulty: 2

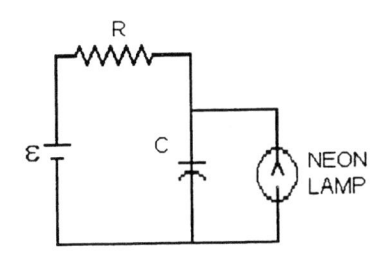

45. A 4-μF capacitor in series with a 5,000-Ω resistor is charged by a 100-V battery. A neon lamp is connected cross the capacitor (in parallel with it). A neon lamp has a very high esistance before it "fires", i.e., ionizes. When the voltage across it reaches 70 V, it fires and its resistance drops almost instantaneously to zero. It then ceases to conduct once the voltage drops below 70 V. This results in periodically discharging the capacitor. This is the kind of circuit used to make roadside warning lights at construction sites. Determine the frequency at which the light will blink in the above circuit.

    Answer: 8.3 Hz Difficulty: 2

MULTIPLE CHOICE

46. What is the unit for the quantity RC?
    a) Ohms
    b) Volt-Ampere/ohm
    c) Seconds
    d) Meters

    Answer: c Difficulty: 2

47. A 2-μF capacitor is charged though a 50,000-Ω resistor. How long does it take for the capacitor to reach 90% of full charge?
    a) 0.9 s
    b) 0.23 s
    c) 2.19 s
    d) 2.3 s

    Answer: b Difficulty: 2

Chapter 17

48. A 4-μF capacitor is charged to 6 V. It is then connected in series with a 3-MΩ resistor and connected to a 12-V battery. How long after being connected to the battery will the voltage across the capacitor be 9 V?
    a) 5.5 s
    b) 8.3 s
    c) 12 s
    d) 16.6 s

    Answer: b Difficulty: 2

49. A 4-MΩ resistor is connected in series with a 0.5-μF capacitor. The capacitor is initially uncharged. The RC combination is charged by a 9-V battery. What is the change in voltage between t = RC and t = 3RC?
    a) 11.40 V
    b) 7.59 V
    c) 5.70 V
    d) 3.80 V

    Answer: d Difficulty: 2

ESSAY

50. A galvanometer has a coil with a resistance of 24 Ω. A current of 180 μA causes full-scale deflection. If the galvanometer is to be used to construct an ammeter that deflects full scale for 10 A, what shunt resistor is required?

    Answer: 43.2 kΩ Difficulty: 2

51. A galvanometer with a coil resistance of 40 Ω deflects full scale for a current of 2 mA. What series resistance should be used with this galvanometer in order to construct a voltmeter that deflects full scale for 50 V?

    Answer: 25,000 Ω Difficulty: 2

MULTIPLE CHOICE

52. An ideal ammeter should
    a) have a high coil resistance.
    b) introduce a very small series resistance into the circuit whose current is to be measured.
    c) introduce a very large series resistance into the circuit whose current is to be measured.
    d) consist of a galvanometer in series with a large resistor.

    Answer: b Difficulty: 2

53. In order to construct a voltmeter from a galvanometer, one normally would
   a) use a very small shunt resistor.
   b) use a very large shunt resistor.
   c) use a very small series resistor.
   d) use a very large series resistor.

   Answer: d Difficulty: 2

54. A galvanometer with a coil resistance of 80 Ω deflects full scale for a current of 2 mA. What series resistance is required to convert it to a voltmeter reading full scale for 200 V?
   a) 0.0008 Ω
   b) 99,920 Ω
   c) 100,000 Ω
   d) 100,080 Ω

   Answer: b Difficulty: 2

55. An unknown resistor is wired in series and an ammeter and a voltmeter are placed in parallel across both the resistor and the ammeter. This network is then placed across a battery. If one computes the value of the resistance by dividing the voltmeter reading by the ammeter reading, the value obtained
   a) is less than the true resistance.
   b) is greater than the true resistance.
   c) is the true resistance.
   d) could be any of the above. It depends on other factors.

   Answer: b Difficulty: 2

56. An unknown resistor is wired in series and an ammeter and a voltmeter are placed in parallel across the resistor only. This network is then connected to a battery. If one computes the value of the resistance by dividing the voltmeter reading by the ammeter reading, the value obtained
   a) is less than the true resistance.
   b) is greater than the true resistance.
   c) is the true resistance.
   d) could be any of the above. It depends on other factors.

   Answer: a Difficulty: 2

57. Increasing the resistance of a voltmeter's series resistance
   a) allows it to measure a larger voltage at full-scale deflection.
   b) allows it to measure a smaller voltage at full-scale deflection.
   c) enables more current to pass through the meter movement at full-scale deflection.
   d) converts it to an ammeter.

   Answer: a Difficulty: 2

Chapter 17

58. A current reading is obtained by properly placing an ammeter in a circuit consisting of one resistor and one battery. As a result,
   a) the voltage drop across the resistor increases.
   b) the current flowing in the circuit increases.
   c) the current flowing in the circuit decreases.
   d) the current flowing in the circuit does not change.

   Answer: c Difficulty: 2

59. A voltage reading is obtained by placing a voltmeter across a resistor. What happens to the total current flowing in the circuit as a result of this action?
   a) The current increases.
   b) The current decreases.
   c) The current does not change.
   d) The current increases if the meter's internal resistance is less than the original resistance in the circuit and decreases if its internal resistance is greater than the circuit's original resistance.

   Answer: a Difficulty: 2

TRUE/FALSE

60. One amp of current is enough to kill a person.

   Answer: T Difficulty: 2

61. Fuses contain a bimetallic strip.

   Answer: F Difficulty: 2

ESSAY

62. Why does a lamp burn out quicker if it is turned on and off more times than one that is left burning?

   Answer:
   The Tungsten filament has a positive temperature coefficient of resistance. Thus when it is first turned on, its current is excessive and the filament for an instant evaporates some of its surface, thereby becoming thinner.
   Difficulty: 2

MULTIPLE CHOICE

63. The minimum current that will usually kill the average person is:
    a) 10. A
    b) 1. A
    c) 0.1 A
    d) 0.01 A
    e) 0.1 mA

    Answer: c Difficulty: 1

64. A metal wire conductor would pass less current if its:
    a) length were decreased
    b) cross-sectional area were larger
    c) conductivity were larger
    d) resistivity were smaller
    e) none of the above

    Answer: e Difficulty: 2

FILL-IN-THE-BLANK

65. An air conditioner uses 2000. watts at 120. volts. What is its current and resistance?

    Answer: 16.7 amperes and 7.20 ohms Difficulty: 2

MULTIPLE CHOICE

66. We pay the "power company" for
    a) voltage
    b) power
    c) current
    d) resistance
    e) energy

    Answer: e Difficulty: 1

67. A copper wire of 1.0 cm^2 cross sectional area would have to how long to have a resistance of 1.0 ohm?
    a) 5.9 m
    b) 5.9 x 10^2 m
    c) 5.9 km
    d) 5.9 x 10^4 m
    e) 5.9 Mm

    Answer: c Difficulty: 2

Chapter 17

68. Which of the following materials has the largest conductivity (at 20C):
a) copper
b) iron
c) silver
d) platinum
e) glass

Answer: c   Difficulty: 1

ESSAY

69. My hair dryer is rated at 1500.watts and my television only uses 100.watts, yet why do I pay more for electricity for my TV than for my hair dryer in a year?

Answer:
One pays for energy, not power. I use my TV so much longer that its energy usage (powerXtime) is more than the hair dryer.
Difficulty: 2

FILL-IN-THE-BLANK

70. An Tungsten wire is 1.50 m long and has a diameter of 1.00 mm. At 20.°C a current of 50.mA flows through the wire.
a) Determine the potential across the wire.
b) If the wire temperature increases by 100.C°, what potential is now required to produce a current of 50.mA ?
[ρ=5.6x10^{-8} ohm-m; α=0.0045(C°)$^{-1}$]

Answer: a) 5.4 mV   b) 7.8 mV   Difficulty: 2

MULTIPLE CHOICE

71. In 5.0 minutes, 500. Coulombs pass through a 24.ohm resistor. What is the potential across the resistor?
a) 35.v
b) 30.v
c) 40.v
d) 300.v
e) 400.v

Answer: c   Difficulty: 2

ESSAY

72. Why has superconductivity not been used extensively in "everyday products"?

    Answer:
    We have so far been able to create superconductivity at very low temperatures (approximately $-150.°C$). Expensive refrigeration techniques keep superconductivity expensive and complicated.
    Difficulty: 2

FILL-IN-THE-BLANK

73. In order for the human body to conduct 100.mA or more, connected to 120.volts, its resistance would have to be less than what?

    Answer: $1.20 \, K\Omega$ Difficulty: 2

MULTIPLE CHOICE

74. An insulator (like glass) has about how many orders of magnitude larger resistivity than a metal (like copper)?
    a) 3
    b) 10
    c) 20
    d) 10^{10}
    e) 10^{20}

    Answer: c   Difficulty: 1

ESSAY

75. Why can a 25.watt flourescent lamp produce as much illumination as a 100.watt incandescent lamp?

    Answer:
    The incandescent lamp produces heat out of the majority of the input energy and only a minority fraction becomes light. The flourescent lamp produces more light, less heat.
    Difficulty: 2

Chapter 17

FILL-IN-THE-BLANK

76. The label on a "rest room" hand dryer reads 20.A at 120.volts.
    a) How many Joules does it use when it cycles on for 30.seconds? ___
    b) At 6.cents/Kw-h, how much does a "drying" cost? _____

    Answer: a) 20A(120V)30s = 72.KJ   b) 0.12 cents   Difficulty: 2

MULTIPLE CHOICE

77. A 200.cm long Aluminum wire and a 300.cm long Copper wire have similar diameters at 20.°C. At what temperature are the two resistances the same?
    Data: Al 2.82×10^{-8} Ω-m , $\alpha = 4.29 \times 10^{-3}$ (C°)$^{-1}$
          Cu 1.70×10^{-8} Ω-m , $\alpha = 6.80 \times 10^{-3}$ (C°)$^{-1}$

    a) 32.°C
    b) 42.°C
    c) 52.°C
    d) 62.°C
    e) 72.°C

    Answer: e   Difficulty: 2

ESSAY

78. Usually one can touch the terminals of a battery (e.g. 1.5 volts) with no dire consequences. Why do hospital intensive care rooms go to great effort and expense to attach "ground wires" to equipment in order to guard against the presence of low potentials? (i.e. why are low voltages hazardous in hospitals?)

    Answer:
    Most of the body resistance is at the relatively dry skin-air boundary and this is sufficient to prevent dangerous currents through the body. In a hospital setting one may have needles or other probes penetrating this protective layer and the potential exists for much larger currents being driven by relatively small voltages.
    Difficulty: 2

FILL-IN-THE-BLANK

79. My auto battery is rated 84.Ampere-hours at 12.volts.
    A) How long may I run the starter motor which uses 400.A?
    B) How much energy does it store?

    Answer: A) 0.21 hour = 13.minutes   B) 3.6 MJ = 1.0 Kw-h   Difficulty: 2

Chapter 18

MATCHING

Match the unit to the physical quantity.

a) T-m/A
b) A-m^2
c) Wb/m^2
d) dimensionless
e) m^{-1}
f) A-m^2-T^2

1. permeability

    Answer: a) T-m/A Difficulty: 2

2. magnetic moment

    Answer: b) A-m^2 Difficulty: 2

3. magnetic field

    Answer: c) Wb/m^2 Difficulty: 2

4. relative permeability

    Answer: d) dimensionless Difficulty: 2

5. turns per meter

    Answer: e) m^{-1} Difficulty: 2

6. torque

    Answer: f) A-m^2-T^2 Difficulty: 2

TRUE/FALSE

7. Breaking a bar magnet in half will produce a north magnetic monopole and a south magnetic monopole.

    Answer: F Difficulty: 2

ESSAY

8. A proton moving at 5 x 10^4 m/s horizontally enters a region where a magnetic field of 0.12 T is present, directed vertically downward. What force acts on the proton?

    Answer: 9.6 x 10^{-16} N Difficulty: 2

Chapter 18

9. An electron traveling due north with speed 4×10^5 m/s enters a region where the earth's magnetic field has the magnitude 5×10^{-5} T and is directed downward at 45° below horizontal. What force acts on the electron?

   Answer: 2.26×10^{-18} N Difficulty: 2

10. Two long parallel wires carry currents of 10 A in opposite directions. They are separated by 40 cm. What is the magnetic field in the plane of the wires at a point that is 20 cm from one wire and 60 cm from the other?

    Answer: 6.67×10^{-18} N Difficulty: 2

11. What is the magnetic field at the center of a circular loop of wire of radius 4 cm when a current of 2 A flows in the wire?

    Answer: 3.14×10^{-5} N Difficulty: 2

12. A very long straight wire carrying 2 A lies along the x-axis. A second very long parallel wire carrying 3 A in the same direction is positioned parallel to the x-axis and passes through the point (0, 1m). What is the magnitude of the magnetic field at the point (0, 0, 1m)?

    Answer: 7.62×10^{-7} N Difficulty: 2

13. A proton is accelerated from rest through 500 V. It enters a magnetic field of 0.30 T oriented perpendicular to its direction of motion. Determine (a) the radius of the path it follows (b) the frequency with which it moves around this path.

    Answer: (a) 0.011 cm   (b) 4.57 MHz Difficulty: 2

14. A deuteron has the same charge as an electron but approximately twice the proton's mass. Suppose deuterons are accelerated in a cyclotron in an orbit of 75 cm at a frequency of 8 MHz. What magnetic field is needed?

    Answer: 1.07 T Difficulty: 2

MULTIPLE CHOICE

15. The SI unit of magnetic field is the
    a) weber.
    b) gauss.
    c) tesla.
    d) lorentz.

    Answer: c Difficulty: 2

16. A vertical wire carries a current straight up in a region where the magnetic field vector points due north. What is the direction of the resulting force on this current?
   a) Down
   b) North
   c) East
   d) West

   Answer: d Difficulty: 2

17. A charged particle is injected into a uniform magnetic field such that its velocity vector is perpendicular to the magnetic field vector. Ignoring the particle's weight, the particle will
   a) move in a straight line.
   b) follow a spiral path.
   c) move along a parabolic path.
   d) follow a circular path.

   Answer: d Difficulty: 2

18. A charged particle is observed traveling in a circular path in a uniform magnetic field. If the particle had been traveling twice as fast, the radius of the circular path would be
   a) twice the original radius.
   b) four times the original radius.
   c) one-half the original radius.
   d) one-fourth the original radius.

   Answer: a Difficulty: 2

19. A particle carrying a charge of +1 e travels in a circular path in a uniform magnetic field. If instead the particle carried a charge of +2 e, the radius of the circular path would have been
   a) twice the original radius.
   b) four times the original radius.
   c) one-half the original radius.
   d) one-fourth the original radius.

   Answer: c Difficulty: 2

20. At double the distance from a long current-carrying wire, the strength of the magnetic field produced by that wire decreases to
   a) 1/8 of its original value.
   b) 1/4 of its original value.
   c) 1/2 of its original value.
   d) none of the above.

   Answer: c Difficulty: 2

Chapter 18

21. A horizontal wire carries a current straight toward you. From your point of view, the magnetic field caused by this current
    a) points directly away from you.
    b) points to the left.
    c) circles the wire in a clockwise direction.
    d) circles the wire in a counter-clockwise direction.

    Answer: d   Difficulty: 2

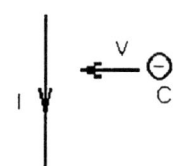

22. A wire lying in the plane of the page carries a current toward the bottom of the page, as shown. What is the direction of the magnetic force it produces on an electron that is moving perpendicularly toward the wire, also in the plane of the page, from your right?
    a) Zero
    b) Perpendicular to the page and towards you
    c) Perpendicular to the page and away from you
    d) Toward the top of the page
    e) Toward the bottom of the page

    Answer: e   Difficulty: 2

23. Consider two current-carrying circular loops. Both are made from one strand of wire and both carry the same current, but one has twice the radius of the other. Compared to the magnetic field at the center of the smaller loop, the magnetic field at the center of the larger loop is
    a) 8 times stronger.
    b) 4 times stronger.
    c) 2 times stronger.
    d) none of the above.

    Answer: d   Difficulty: 2

TRUE/FALSE

24. Parallel wires carrying currents in the same direction will repel each other.

    Answer: T   Difficulty: 2

MULTIPLE CHOICE

25. What is the force per meter on a straight wire carrying 5 A when it is placed in a magnetic field of 0.02 T? The wire makes an angle of 27° with respect to the magnetic field lines.
    a) 0.022 N
    b) 0.045 N
    c) 0.17 N
    d) 0.26 N

    Answer: b Difficulty: 2

26. A solenoid 20 cm long is wound with 5000 turns of wire. What magnetic field is produced at the center of the solenoid when a current of 10 A flows?
    a) 1.6 T
    b) 0.84 T
    c) 0.67 T
    d) 0.31 T

    Answer: d Difficulty: 2

TRUE/FALSE

27. When a ferromagnetic material is placed in an external magnetic field, the net magnetic field of its magnetic domain becomes smaller.

    Answer: F Difficulty: 2

ESSAY

28. What is the magnetic moment of a rectangular loop of 120 turns that carries 6 A if its dimensions are 4 cm x 8 cm?

    Answer: 2.30 A-m^2 Difficulty: 2

29. Four long parallel wires each carry 2 A in the same direction. They are parallel to the z-axis, and they pass through the corners of a square of side 4 cm positioned in the x-y plane. What force does one of the wires experience due to the other wires?

    Answer: 2.12 x 10^{-5} N Difficulty: 2

30. A circular loop of wire of cross-sectional area 0.12 m^2 consists of 200 turns, each carrying 0.5 A. It is placed in a magnetic field of 0.05 T oriented at 30° to the plane of the loop. What torque acts on the loop?

    Answer: 0.52 N-m Difficulty: 2

Chapter 18

MULTIPLE CHOICE

31. A charged particle moves with a constant speed through a region where a uniform magnetic field is present. If the magnetic field points straight upward, the magnetic force acting on this particle will be maximum when the particle moves
    a) straight upward.
    b) straight downward.
    c) upward at an angle of 45° above the horizontal.
    d) none of the above.

    Answer: d Difficulty: 2

32. At a particular instant, a proton moves eastward at speed V in a uniform magnetic field that is directed straight downward. The magnetic force that acts on it is
    a) zero.
    b) directed upward.
    c) to the south.
    d) none of the above.

    Answer: d Difficulty: 2

33. At a particular instant, an electron moves eastward at speed V in a uniform magnetic field that is directed straight downward. The magnetic force that acts on it is
    a) zero.
    b) directed upward.
    c) directed to the south.
    d) none of the above.

    Answer: c Difficulty: 2

34. A charged particle moves and experiences no magnetic force. From this we can conclude that
    a) no magnetic field exists in that region of space.
    b) the particle is moving parallel to the magnetic field.
    c) the particle is moving at right angles to the magnetic field.
    d) either no magnetic field exists or the particle is moving parallel to the field.

    Answer: d Difficulty: 2

35. Consider two current-carrying circular loops. Both are made from one strand of wire and both carry the same current, but one has twice the radius of the other. Compared to the magnetic moment of the smaller loop, the magnetic moment of the larger loop is
   a) 16 times stronger.
   b) 8 times stronger.
   c) 4 times stronger.
   d) 2 times stronger.

   Answer: c   Difficulty: 2

TRUE/FALSE

36. A charged particle moving along a direction perpendicular to the plane of a loop of current-carrying wire, and which passes through the loop's center, will experience no magnetic force.

   Answer: T   Difficulty: 2

MULTIPLE CHOICE

37. What fundamental fact underlies the operation of essentially all electric motors?
   a) Opposite electric charges attract and like charges repel.
   b) A current-carrying conductor placed perpendicular to a magnetic field will experience a force.
   c) Alternating current and direct current are both capable of doing work.
   d) Iron is the only element that is magnetic.
   e) A magnetic north pole carries a positive electric charge, and a magnetic south pole carries a negative electric charge.

   Answer: b   Difficulty: 2

TRUE/FALSE

38. A motor is a device that converts mechanical energy into electrical energy.

   Answer: F   Difficulty: 2

39. A split-ring commutator in a DC motor reverses the polarity and current each half-cycle.

   Answer: T   Difficulty: 2

40. The torque on the rotating loop of wire in a DC motor is constant, in magnitude, as the loop rotates.

   Answer: F   Difficulty: 2

Chapter 18

MULTIPLE CHOICE

41. In a pair of accelerating plates, such as found inside a CRT, the electrons are emitted
    a) from the cathode which is positive, toward the anode which is positive.
    b) from the cathode which is negative, toward the anode which is positive.
    c) from the anode which is positive, toward the cathode which is negative.
    d) from the anode which is negative, toward the cathode which is positive.

    Answer: b Difficulty: 2

ESSAY

42. In a mass spectrometer a particle of mass m and charge q is accelerated through a potential difference V and allowed to enter a magnetic field B, where it is deflected in a semi-circular path of radius R. The magnetic field is uniform and oriented perpendicular to the velocity of the particle. Derive an expression for the mass of the particle in terms of B, q, V, and R.

    Answer: $m = qB^2 R^2 /2V$ Difficulty: 2

MULTIPLE CHOICE

43. A beam of electrons is accelerated through a potential difference of 10 kV before entering a velocity selector. If the B-field of the velocity selector has a value of 10^{-2} T, what value of the E-field is required if the particles are to be undeflected?
    a) 2.26×10^3 V/m
    b) 5.93×10^5 V/m
    c) 6.04×10^5 V/m
    d) 7.24×10^6 V/m

    Answer: b Difficulty: 2

44. What happens to the cyclotron frequency of a charged particle if its speed doubles?
    a) It triples.
    b) It doubles.
    c) It halves.
    d) It does not change.

    Answer: d Difficulty: 2

45. A cyclotron operates at 10 MHz. What magnetic field is needed to accelerate protons?
   a) 0.12 T
   b) 0.25 T
   c) 0.66 T
   d) 1.12 T

   Answer: c   Difficulty: 2

46. A doubly charged ion with velocity 6×10^6 m/s moves in a path of radius 30 cm in a magnetic field of 0.8 T in a mass spectrometer. What is the mass of this ion?
   a) 12.8×10^{-27} kg
   b) 6.68×10^{-27} kg
   c) 3.34×10^{-27} kg
   d) 8.22×10^{-27} kg

   Answer: a   Difficulty: 2

ESSAY

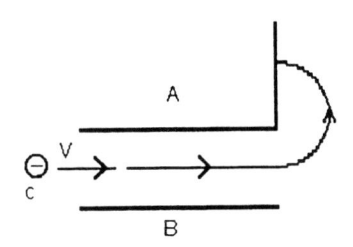

47. An electron enters the velocity selector of a mass spectrometer in the direction shown in the sketch. If it emerges along the path shown, which plate (A or B) has the higher potential?

   Answer: Plate B   Difficulty: 2

TRUE/FALSE

48. The earth's magnetic field at its magnetic equator is approximately 10^4 T.

   Answer: F   Difficulty: 2

MULTIPLE CHOICE

49. The source of the earth's magnetic field is thought to be due to the motion of charged particles in the earth's
   a) mantle.
   b) crust.
   c) inner core.
   d) outer core.

   Answer: d   Difficulty: 2

Chapter 18

TRUE/FALSE

50. "Magnetic bottles" are composed of non-uniform magnetic field lines.

    Answer: T Difficulty: 2

MULTIPLE CHOICE

51. A proton with velocity 2×10^5 m/s enters a region where a uniform magnetic field of 0.3 T is present. The proton's velocity makes an angle of 60° with the magnetic field. What is the pitch of the helical path followed by the proton?

    (The "pitch" is the distance traveled parallel to the magnetic field while making one complete revolution around a magnetic field line.)
    a) 0.022 m
    b) 0.077 m
    c) 0.140 m
    d) 0.250 m

    Answer: a Difficulty: 2

ESSAY

52. How much would the electricity cost to leave a 10.watt night light on every day and night for a week assuming an electricity rate of 6.0 cents per kw-h ?

    $ _____

    Answer: $ 0.10 Difficulty: 2

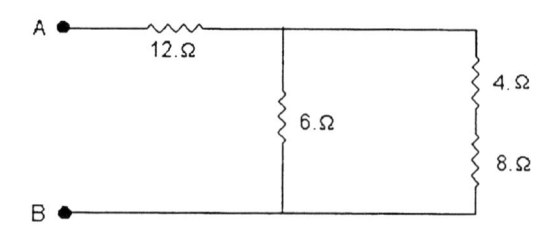

53. Find the equivalent resistance of the given network between points A and B:
    ans._____

    Answer: 16.ohms Difficulty: 2

54. Why do power tools with a plastic case not have a 3 prong plug but metal encased tools have (or should have) 3 prong plugs?

    Answer:
    The third wire (large prong plug) connects to ground and grounds the metal case. A hot wire accidentally coming in contact with the case will blow a fuse. A plastic case is insulation and does not need to be grounded for safety.
    Difficulty: 1

TRUE/FALSE

55. In a polarized plug, the large slit connects to the hot side and the small slit connects to the neutral, or ground, side.

    Answer: F Difficulty: 1

ESSAY

56. When double checking a wire which is assumed to be not "hot", you might see an electrician "testing" for electricity by briefly touching the suspicious wire WITH THE BACK OF THE HAND. Why do you suppose the front of the fingers are not used?

    Answer:
    If current flows through the fingers the muscles are likely to contract. If the front of the hand were used then the hand might be clamped to the wire but using the back of the hand prevents this worsening of the situation.
    Difficulty: 2

FILL-IN-THE-BLANK

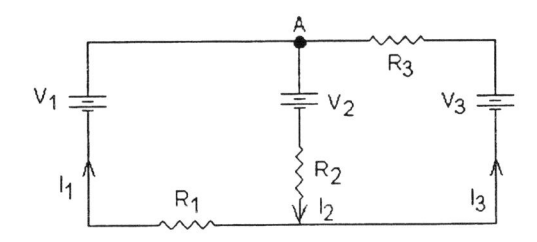

57. Write the Kirchhoff loop equation for the entire outside loop (see figure).

    Answer: $V_1 + I_3 R_3 + V_3 - I_1 R_1 = 0$ Difficulty: 2

Chapter 18

58. For the circuit illustrated in the figure, write the Kirchhoff current equation for the node labeled A.

    Answer: $I_1 - I_2 + I_3 = 0$ Difficulty: 2

MULTIPLE CHOICE

59. The usual household fuse or circuit breaker is rated at
    a) 2 A
    b) 20 A
    c) 2 mA
    d) 20 mA
    e) 200 A
    f) 200 mA

    Answer: b   Difficulty: 1

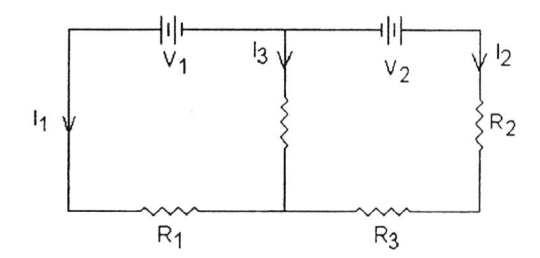

60. In the accompanying figure, Which of the following relations is true:
    a) $I_1 + I_2 + I_3 = 0$
    b) $I_1 - I_2 - I_3 = 0$
    c) $I_1 - I_2 + I_3 = 0$
    d) $-I_1 + I_2 + I_3 = 0$
    e) $I_1 + I_2 - I_3 = 0$

    Answer: a   Difficulty: 2

61. In the figure, which of the following equations is true:
    a) $V_1 - I_1 R_1 - I_2 R_2 + V_2 - I_2 R_3 = 0$
    b) $V_1 - I_1 R_1 + I_2 R_2 - V_2 + I_2 R_3 = 0$
    c) $V_1 + I_1 R_1 + I_2 R_2 + V_2 + I_2 R_3 = 0$
    d) $V_1 + I_1 R_1 - I_2 R_2 - V_2 - I_3 R_4 = 0$

    Answer: b   Difficulty: 2

FILL-IN-THE-BLANK

62. If the internal resistance of my 1.5 volt battery is 0.50 ohms, what is the maximum possible power and current that can be delivered from it?

    Answer: 4.5 watts at 3.0 Amperes   Difficulty: 2

MULTIPLE CHOICE

63. The internal resistance of a typical "12.v" auto battery is MORE or LESS than 1.ohm?
    a) MORE
    b) LESS
    c) none of the above

    Answer: b Difficulty: 2

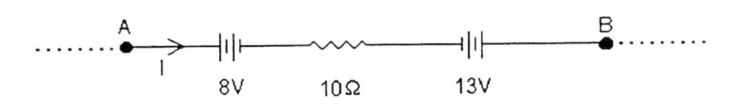

64. Refer to the portion of a circuit given in the figure. What is the potential difference $V_A - V_B$ if $I = 5.0$ Amperes?
    a) 71.v
    b) 63.v
    c) 55.v
    d) 45.v
    e) 35.v
    f) 5.v

    Answer: d Difficulty: 2

65. A 1234.pF capacitor and a 5.6×10^6 ohm resistor are connected in series to 78.volts EMF. Approximately how long does it take the capcitor to become almost fully charged?
    a) 1. s
    b) 0.01 s
    c) 0.1 ms
    d) 1. µs
    e) 0.01 µs
    f) 1. ps

    Answer: b Difficulty: 2

ESSAY

66. When charging a capacitor with a battery, how would the internal resistance of the battery affect the charging of the capacitor?

    Answer:
    Given enough charging time, the final potential on the capacitor will still equal the emf of the battery. However the charging rate will be influenced: the battery resistance adds to the circuit resistance in the time constant.

    Difficulty: 2

Chapter 18

MULTIPLE CHOICE

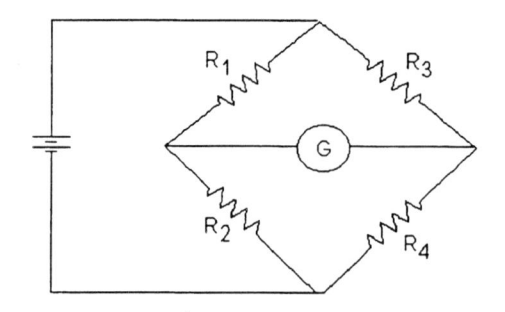

67. In a wheatstone bridge (see figure) the resistance R_1 is varied until no current flows through the Galvanometer (G). Then one knows:
    a) $R_1 R_2 = R_3 R_4$
    b) $R_1 / R_4 = R_3 / R_2$
    c) $R_1 + R_2 = R_3 + R_4$
    d) $R_1 R_4 = R_2 R_3$

    Answer: d Difficulty: 2

FILL-IN-THE-BLANK

68. Two resistors in series equal 9.0 ohms total; but the same two resistors in parallel are equivalent to 2.0 ohms. What are the resistor values?

    Answer: 3.0 and 6.0 ohms Difficulty: 2

Chapter 19

MATCHING

Match the unit or quantity to the physical quantity.

a) N/m²   d) deci-meters   g) tesla
b) 10^{-7} meters   e) weber   h) one meter
c) volts   f) 10^{-11} meters

1. radiation pressure

    Answer: a) N/m² Difficulty: 2

2. typical wavelength of visible light

    Answer: b) 10^{-7} meters Difficulty: 2

3. back emf

    Answer: c) volts Difficulty: 2

4. typical wavelength of microwaves

    Answer: d) deci-meters Difficulty: 2

5. magnetic flux

    Answer: e) weber Difficulty: 2

6. typical wavelength of X-rays

    Answer: f) 10^{-11} meters Difficulty: 2

7. magnetic field

    Answer: g) tesla Difficulty: 2

8. typical wavelength of TV waves

    Answer: h) one meter Difficulty: 2

Chapter 19

MULTIPLE CHOICE

9. All of the following are units of magnetic flux except:
   a) T-m^2
   b) T/volt-m
   c) weber
   d) volt-second

   Answer: b Difficulty: 2

10. Magnetic flux density has the same units as
    a) magnetic flux.
    b) back emf.
    c) magnetic dipole.
    d) magnetic field.

    Answer: d Difficulty: 2

ESSAY

11. A coil of 40 turns and cross-sectional area 12 cm^2 is oriented perpendicular to a magnetic field, which varies from zero to 1.2 T in 0.02 s. What emf is induced in the coil?

    Answer: 2.88 V Difficulty: 2

12. A circular loop of one turn and radius 5 cm is positioned with its axis parallel to a magnetic field of 0.6 T. By means of high explosives, the area of the loop is suddenly reduced to essentially zero in 0.5 ms. What emf is induced in the loop?

    Answer: 9.4 V Difficulty: 2

13. An eagle, with a wingspread of 2 m, flies due north at 8 m/s in a region where the vertical component of the earth's magnetic field is 0.2 x 10^{-4} T. What emf would be developed between the eagle's wing tips? (It has been speculated that this phenomenon could play a role in the navigation of birds, but the effect is too small, in all likelihood.)

    Answer: 0.32 mV Difficulty: 2

14. A coil of 160 turns and area 0.20 m^2 is placed with its axis parallel to a magnetic field of 0.4 T. The magnetic field changes from 0.4 T in the x-direction to 0.4 T in the negative x-direction in 2 s. If the resistance of the coil is 16 Ω, at what rate is power generated in the coil?

    Answer: 10.2 W Difficulty: 2

MULTIPLE CHOICE

15. A wire moves across a magnetic field. The emf produced in the wire depends on
    a) the field's magnetic flux.
    b) the length of the wire.
    c) the orientation of the wire with respect to the magnetic field vector.
    d) all of the above.

    Answer: d Difficulty: 2

16. Faraday's law of induction states that the emf induced in a loop of wire is proportional to
    a) the magnetic flux.
    b) the magnetic flux density times the loop's area.
    c) the time variation of the magnetic flux.
    d) current divided by time.

    Answer: c Difficulty: 2

17. Doubling the number of loops of wire in a coil produces what kind of change on the induced emf, assuming all other factors remain constant?
    a) The induced emf is 4 times as much.
    b) The induced emf is 3 times as much.
    c) The induced emf is twice as much.
    d) There is no change in the induced emf.

    Answer: c Difficulty: 2

18. The flux through a coil changes from 4×10^{-5} Wb to 5×10^{-5} Wb in 0.1 s. What emf is induced in this coil?
    a) 5×10^{-4} V
    b) 4×10^{-4} V
    c) 1×10^{-4} V
    d) None of the above

    Answer: c Difficulty: 2

19. A flux of 4×10^{-5} Wb is maintained through a coil for 0.5 s. What emf is induced in this coil by this flux?
    a) 8×10^{-5} V
    b) 4×10^{-5} V
    c) 2×10^{-5} V
    d) No emf is induced in this coil.

    Answer: d Difficulty: 2

Chapter 19

20. A coil lies flat on a table top in a region where the magnetic field vector points straight up. The magnetic field vanishes suddenly. When viewed from above, what is the sense of the induced current in this coil as the field fades?
    a) The induced current flows counter-clockwise.
    b) The induced current flows clockwise.
    c) There is no induced current in this coil.
    d) The current flows clockwise initially, and then it flows counter-clockwise before stopping.

    Answer: a Difficulty: 2

21. A circular loop of wire is rotated at constant angular speed about an axis whose direction can be varied. In a region where a uniform magnetic field points straight down, what must be the orientation of the loop's axis of rotation if the induced emf is to be zero?
    a) Any horizontal orientation will do.
    b) It must make an angle of 45° to the vertical.
    c) It must be vertical.
    d) None of the above.

    Answer: a Difficulty: 2

22. A horizontal metal bar rotates at a constant angular velocity w about a vertical axis through one of its ends while in a constant magnetic field B that is directed down. The emf induced between the two ends of the bar is
    a) constant and proportional to the product $B \ell$.
    b) constant and proportional to the product $B \ell^2$.
    c) constant and proportional to the product $B^2 \ell^2$.
    d) none of the above.

    Answer: a Difficulty: 2

23. A horizontal metal bar that is 2 m long rotates at a constant angular velocity of 2 rad/s about a vertical axis through one of its ends while in a constant magnetic field of 5×10^{-5} T. If the magnetic field vector points straight down, what emf is induced between the two ends of the bar?
    a) 5.6×10^{-3} V
    b) 3×10^{-3} V
    c) 2×10^{-4} V
    d) 1.6×10^{-4} V

    Answer: c Difficulty: 2

24. Consider a circular ring of material through which some magnetic field lines pass. In order to induce an emf around the ring, it is necessary that
   a) the material be a conductor.
   b) the material be an insulator.
   c) the number of magnetic field lines passing through the ring be increased.
   d) the ring be rotated.
   e) none of the above statements is true.

   Answer: e Difficulty: 2

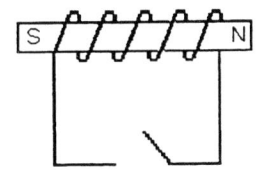

25. A bar magnet is positioned inside a coil. In the following, "work" refers to any work done as a consequence of the fact that the bar is magnetic. If the bar is suddenly pulled out of the coil,
   a) no work will be done, independent of whether or not the switch is closed.
   b) more work will be done if the switch is open.
   c) more work will be done if the switch is closed.
   d) equal (non-zero) amounts of work will be done whether or not the switch is open.
   e) whether the work done is positive or negative depends on whether the magnet is pulled out the right or left end of the coil.

   Answer: c Difficulty: 2

Chapter 19

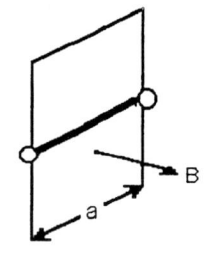

26. A connecting wire of negligible resistance is bent into the shape shown here. The long portions of the wire are vertical. A bar of length a, of resistance R, and mass m can slide downward without friction on the vertical sections. A uniform horizontal magnetic field B is present. The cross-bar is released from rest, and allowed to fall under the influence of gravity. The vertical wires are long enough so that their length is of no concern. Under these conditions, how will the falling bar behave?
a) It will continue to accelerate indefinitely.
b) It will slow down and finally come to rest in equilibrium.
c) It will reach a constant terminal velocity of $mgR/2a^2B^2$.
d) It will reach a constant terminal velocity of mgR/a^2B^2.
e) It will reach a constant terminal velocity of $2mgR/a^2B^2$.

Answer: e   Difficulty: 2

ESSAY

27. A 60-Hz AC generator produces a maximum emf of 170 V. What is the value of the emf 0.025 s after it has its maximum value?

Answer: 85 V   Difficulty: 2

28. The windings of a DC motor have a resistance of 6 Ω. The motor operates on 120 V AC, and when running at full speed it generates a back emf of 105 V. (a) What is the starting current of the motor? (b) What current does the motor draw when operating at full speed?

Answer: (a) 20 A   (b) 2.5 A   Difficulty: 2

29. Suppose that you wish to construct a simple AC generator with an output of 12 V maximum when rotated at 60 Hz. A magnetic field of 0.05 T is available. If the area of the rotating coil is 100 cm², how many turns are needed?

Answer: 64 turns   Difficulty: 2

MULTIPLE CHOICE

30. A circular loop of wire is rotated at constant angular speed about an axis whose direction can be varied. In a region where a uniform magnetic field points straight down, what must be the orientation of the loop's axis of rotation if the induced emf is to be a maximum?
    a) Any horizontal orientation will do.
    b) It must make an angle of 45° to the vertical.
    c) It must be vertical.
    d) None of the above.

    Answer: a Difficulty: 2

31. A generator coil rotates through one revolution 60 times each second. The frequency of the emf is
    a) 30 Hz.
    b) 60 Hz.
    c) 120 Hz.
    d) cannot be determined from the information given.

    Answer: b Difficulty: 2

32. An AC generator has 80 rectangular loops on its armature. Each loop is 12 cm long and 8 cm wide. The armature rotates at 1200 rpm about an axis parallel to the long side. If the loop rotates in a uniform magnetic field of 0.3 T, which is perpendicular to the axis of rotation, what will be the maximum output voltage of this generator?
    a) 124 V
    b) 167 V
    c) 182 V
    d) 220 V

    Answer: c Difficulty: 2

33. The coil of a generator has 50 loops and a cross-sectional area of 0.25 m². What is the maximum emf generated by this generator if it is spinning with an angular velocity of 4 rad/s in a 2 T magnetic field?
    a) 100 V
    b) 50 V
    c) 400 V
    d) None of the above

    Answer: a Difficulty: 2

Chapter 19

34. The cross-sectional area of an adjustable single loop is reduced from 1 m² to 0.25 m² in 0.1 s. What is the average emf that is induced in this coil if it is in a region where B = 2 T upward, and the coil's plane is perpendicular to B?
    a) 12 V
    b) 15 V
    c) 18 V
    d) 21 V

    Answer: b Difficulty: 2

35. A lightbulb is plugged into a household outlet, and one of the wires leading to it is wound into a coil. Now a slug of iron is slid into the coil. What effect will this have?
    a) The lamp will get brighter because there will be an induced emf that will drive more current through the lamp.
    b) The lamp will dim because energy will now be dissipated in the iron.
    c) The lamp will dim because a back emf will be produced by the coil, and this will reduce the current flowing in the coil and the lamp.
    d) This will have no effect on the brightness of the lamp.
    e) This will not affect the lamp's brightness because the coil will shift the phase, but not the magnitude, of the current through the lamp.

    Answer: c Difficulty: 2

ESSAY

36. An ideal transformer has 60 turns on its primary coil and 300 turns on its secondary coil. If 120 V at 2 A is applied to the primary, what voltage and current are present in the secondary?

    Answer: 600 V; 0.4 A Difficulty: 2

37. A step-down transformer is needed to reduce a primary voltage of 120 V AC to 6 V AC. What turns ratio is required?

    Answer: 20:1 Difficulty: 2

MULTIPLE CHOICE

38. A transformer is a device that
    a) operates on either DC or AC.
    b) operates only on AC.
    c) operates only on DC.
    d) increases the power level of a circuit.

    Answer: b Difficulty: 2

39. The secondary coil of a neon sign transformer provides 7500 V at 0.01 A. The primary coil operates on 120 V. What does the primary draw?
   a) 0.625 A
   b) 6.25 x 10^{-4} A
   c) 0.16 A
   d) 1.66 A

   Answer: a   Difficulty: 2

ESSAY

40. A power transmission line 50 km long has a total resistance of 0.60 Ω. A generator produces 100 V at 70 A. In order to reduce energy loss due to heating of the transmission line, the voltage is stepped up with a transformer with a turns ratio of 100:1. (a) What percentage of the original energy would be lost if the transformer were not used? (b) What percentage of the original energy is lost when the transformer is used?

   Answer: (a) 42%   (b) 0.0042%   Difficulty: 2

MULTIPLE CHOICE

41. The output of a generator is 440 V at 20 A. It is to be transmitted on a line with resistance of 0.6 Ω. To what voltage must the generator output be stepped up with a transformer if the power loss in transmission is not to exceed 0.01% of the original power?
   a) 4.4 kV
   b) 22 kV
   c) 44.5 kV
   d) 72.7 kV

   Answer: d   Difficulty: 2

42. Alternating current and voltage (AC), rather than direct current and voltage (DC), are used for domestic electrical power. The reason for this is based on one of the following physical principles. Which one?
   a) When a current flows through a resistance, heat is generated.
   b) When a current flows through a wire, a magnetic field is created.
   c) When the number of magnetic field lines passing through a circuit changes, a voltage is generated in the circuit.
   d) When a current-carrying wire is placed in a magnetic field, it may experience a force if properly oriented.
   e) Opposite electric charges attract each other, and like charges repel each other.

   Answer: c   Difficulty: 2

Chapter 19

43. The coil of a generator rotates through one complete turn in a uniform magnetic field 50 times every second. When this generator is connected to an external load, the current through the load reverses directions
   a) 25 times every second.
   b) 50 times every second.
   c) 100 times every second.
   d) none of the above.

   Answer: c Difficulty: 2

44. A very long solenoid has 1×10^6 turns per meter and a cross-sectional area of 2×10^{-4} m^2. A short, secondary coil with 1×10^3 turns is wound over the solenoid's midpoint. What emf is induced in the secondary coil if the current in the solenoid is changing at a rate of 4 A/s?
   a) 1.01 V
   b) 1.62 V
   c) 2.71 V
   d) 4.44 V

   Answer: a Difficulty: 2

45. Consider electrons flowing southward along a horizontal power line. The magnetic field generated directly below the wire by the current is pointing:
   a) eastward
   b) westward
   c) northward
   d) southward
   e) up
   f) down

   Answer: b Difficulty: 1

46. Two parallel straight wires carry currents in the same direction. The magnetic forces between the wires are such that:
   a) the wires repel each other
   b) the wires attract each other
   c) the wires feel a clockwise torque
   d) they are pulled in the direction of the wires

   Answer: b Difficulty: 1

47. A Tesla is NOT equal to
   a) 10^4 Gauss
   b) 1 N-Coul^{-1}-s-m^{-1}
   c) 1 Weber/m^2
   d) 1 Weber-m^2

   Answer: c Difficulty: 1

FILL-IN-THE-BLANK

48. The Earth's magnetic field at the surface is appoximately how many Gauss?

    Answer: 1/2   Difficulty: 2

MULTIPLE CHOICE

49. Inside an infinitely long ideal solenoid, the magnetic field
    a) increases with distance from the axis
    b) decreases with distance from the axis
    c) is uniform
    d) depends upon the cross sectional area

    Answer: c Difficulty: 1

FILL-IN-THE-BLANK

50. A 14.Coulomb charge is moving toward you at 4800.m/s. When it is 70.meters away from you, what is the magnitude of the magnetic field it produces at YOUR location?

    Answer: Zero. (The field is zero directly in front and behind the charge)

    Difficulty: 1

MULTIPLE CHOICE

51. The pole of the compass which points toward the Earth's SOUTH geographic pole is a
    a) magnetic north pole
    b) magnetic south pole
    c) positive charge pole
    d) magnetic barber pole

    Answer: b Difficulty: 2

FILL-IN-THE-BLANK

52. In order to trap the starship Enterprise, the Klingons build a huge solenoid 10.light years long with a diameter of 2.million kilometers. Every kilometer length of the solenoid contains 100.turns of wire. What magnetic field is produced in the center of the solenoid with a current of 2000.Amperes?

    B = _____ _____ (MKS units)

    This is how many Gauss ? _____

    Answer: 2.51×10^{-4} Tesla = 2.51 Gauss Difficulty: 2

Chapter 19

MULTIPLE CHOICE

53. The magnetic field near a strong permanent bar magnet might be 500.
    a) microGauss
    b) milliGauss
    c) Gauss
    d) Tesla

    Answer: c Difficulty: 2

54. After landing on an unexplored Klingon planet, Spock tests for the direction of the magnetic field by firing a beam of electrons in various directions and by observing:
    Electrons moving upward feel a magnetic force in the NW direction.
    Electrons moving horizontally North are pushed down.
    Electrons moving horizontally South-East are pushed upward.
    He naturally concludes that the magnetic field at this landing site is in which direction?
    a) W
    b) E
    c) SW
    d) SE
    e) NE
    f) up
    g) down

    Answer: c Difficulty: 2

FILL-IN-THE-BLANK

55. Consider a 1000.Gauss magnetic field along the z axis and an proton moving with an instantaneous velocity of $\mathbf{v} = 3\,\mathbf{i} + 4\,\mathbf{k}$ km/s (note its speed is 5.km/s composed of 3.km/s along x and 4.km/s along the z axis). Determine the radius of its spiraling orbit.
    [1.67×10^{-27} kg and 1.6×10^{-19} Coulomb]

    Answer: 0.3 mm radius Difficulty: 2

56. Captain Planet measures the magnetic forces on a proton which is vertically falling at 1996.m/s at the equator (take the Earth's field to be 0.40 Gauss horizontal and pointing northward). What FORCE would she measure?

    _____ Newtons
    _____ direction
    (N,E,S,W,up,down,etc)

    Answer: 1.3×10^{-20} N toward the EAST Difficulty: 2

MULTIPLE CHOICE

57. The magnetic moment of a current loop does not depend upon:
    a) the current
    b) the area
    c) the magnetic field

    Answer: c Difficulty: 1

FILL-IN-THE-BLANK

58. If you find yourself in a locality where the magnetic declination is "15.° East". In which direction is true north in relation to a compass needle?

    Answer:
    "15° East" means the compass is pointing 15° East of true north. So True North would be 15° WEST of the compass direction.
    Difficulty: 2

MULTIPLE CHOICE

59. A proton, moving in a uniform magnetic field, moves in a circle perpendicular to the field. If the proton's speed tripled, its time to go around its circular path would:
    a) triple
    b) double
    c) half
    d) third
    e) remain the same

    Answer: e Difficulty: 2

FILL-IN-THE-BLANK

60. To trap an alien spaceship, the Galactic Council has constructed a giant superconducting current-loop of radius 1.0 light-year. What current would be necessary to produce 2.0 Tesla at the center of the loop?

    Answer: 1.5×10^{22} A Difficulty: 2

MULTIPLE CHOICE

61. Diamagnetic materials have relative permeabilities
    a) slightly less than one
    b) equal to one
    c) slightly larger than one
    d) much larger than one

    Answer: a Difficulty: 2

Chapter 19

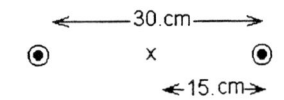

```
        ←――30.cm――→
    ⊙        x        ⊙
         ←15.cm→
```

62. The figure shows two wires, in cross section, carrying 7.0 Amperes out of the page. Determine the total magnetic filed, due to both wires, at point X midway between the wires.
    a) 0.29 Gauss
    b) 0.59 Gauss
    c) 0.15 Gauss
    d) 0.0 Gauss

    Answer: d Difficulty: 2

63. A charged particle moves with its velocity perpendicular to a magnetic field. The field acts to change the particle's:
    a) mass
    b) charge
    c) velocity
    d) magnitude of momentum
    e) energy

    Answer: c Difficulty: 2

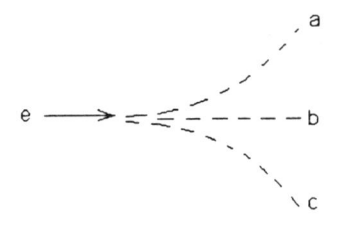

```
                      a
                   ╱
       e ――→  ⇍⇍―――― b
                   ╲
                      c
```

64. Consider a magnetic field pointing out of this page. An electron moving on the page toward the right will (see figure)
    a) curve upward
    b) continue straight ahead
    c) curve downward

    Answer: b Difficulty: 2

65. When a voltage is applied for display on an oscilloscope CRT, it is usually shown:
    a) horizontally
    b) diagonally
    c) vertically

    Answer: c Difficulty: 1

Chapter 20

MATCHING

Match the unit to the physical quantity.

a) s/F
b) dimensionless
c) H/s
d) Z (ohms)
e) henrys
f) farads

1. capacitive reactance

   Answer: a) s/F   Difficulty: 2

2. power factor

   Answer: b) dimensionless   Difficulty: 2

3. inductive reactance

   Answer: c) H/s   Difficulty: 2

4. impedance

   Answer: d) Z (ohms)   Difficulty: 2

5. inductance

   Answer: e) henrys   Difficulty: 2

6. capacitance

   Answer: f) farads   Difficulty: 2

MULTIPLE CHOICE

7. All of the following have the same units except:
   a) inductance.
   b) capacitive reactance.
   c) impedance.
   d) resistance.

   Answer: a   Difficulty: 2

8. The polarity of the voltage in a circuit operating at 60 Hz reverses
   a) 30 times per second.
   b) 60 times per second.
   c) 90 times per second.
   d) 120 times per second.

   Answer: d   Difficulty: 2

Chapter 20

ESSAY

9. A 150-W lamp is placed into a 120-V AC outlet. (a) What are the peak and rms currents? (b) What is the resistance of the lamp?

    Answer: (a) 1.77 A; 1.25 A   (b) 96 Ω   Difficulty: 2

10. The current through a 50-Ω resistor is I = 0.8 sin (240 t). (a) What is the peak current? (b) What is the rms current? (c) At what frequency does the current vary? (d) How much power is dissipated in the resistor?

    Answer: (a) 0.8 A   (b) 0.57 A   (c) 38 Hz   (d) 16 W   Difficulty: 2

11. The peak output of a generator is 20 A and 240 V. What average power is provided by the generator?

    Answer: 2400 W   Difficulty: 2

12. The output voltage of a power supply is V = V_o sin (120 t). At t = 0.01 s the output voltage is 4.2 V. What is the rms voltage of the generator?

    Answer: 3.19 V   Difficulty: 2

MULTIPLE CHOICE

13. What do the letters "rms" stand for?
    a) Real modulated signal
    b) Rationalized metric system
    c) Reduced mean sin (function)
    d) Root mean square

    Answer: d   Difficulty: 2

14. The amount of AC current that produces the same average joule heating effect as 1 ampere of DC current is
    a) 1 A peak current.
    b) 1 A peak-to-peak current.
    c) $1/\sqrt{2}$ A peak current.
    d) $\sqrt{2}$ A peak current.

    Answer: d   Difficulty: 2

TRUE/FALSE

15. In a purely inductive AC circuit, the current lags behind the voltage by 90°.

    Answer: T   Difficulty: 2

ESSAY

16. At what frequency does a 10-μF capacitor have a reactance of 1200 Ω?

    Answer: 13.3 Hz Difficulty: 2

MULTIPLE CHOICE

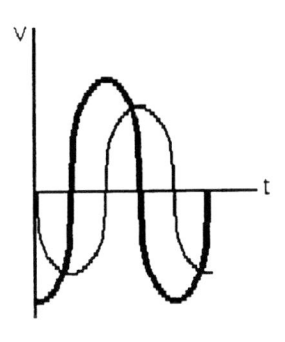

17. Compare the voltage waves from two different AC circuits. The absolute value of the phase difference is
    a) zero
    b) 45°
    c) 90°
    d) 180°

    Answer: b Difficulty: 2

ESSAY

18. What is the reactance of a 1-mH inductor at 60 Hz?

    Answer: 0.0063 Ω Difficulty: 2

19. What current flows in a 60-mH inductor when 120 V AC at a frequency of 20 kHz is applied to it?

    Answer: 15.9 mA Difficulty: 2

20. What capacitance will have the same reactance as a 100-mH inductance in a 120-V, 60-Hz circuit?

    Answer: 70 μF Difficulty: 2

Chapter 20

MULTIPLE CHOICE

21. A resistor and a capacitor are connected in series to an ideal battery of constant terminal voltage. At the moment contact is made with the battery, the voltage across the resistor is
    a) greater than the battery's terminal voltage.
    b) less than the battery's terminal voltage, but greater than zero.
    c) equal to the battery's terminal voltage.
    d) zero.

    Answer: c Difficulty: 2

22. A resistor and a capacitor are connected in series to an ideal battery of constant terminal voltage. At the moment contact is made with the battery, the voltage across the capacitor is
    a) greater than the battery's terminal voltage.
    b) less than the battery's terminal voltage, but greater than zero.
    c) equal to the battery's terminal voltage.
    d) zero.

    Answer: d Difficulty: 2

23. A resistor and a capacitor are connected in series to an ideal battery of constant terminal voltage. When this system reaches its steady-state, the voltage across the resistor is
    a) greater than the battery's terminal voltage.
    b) less than the battery's terminal voltage, but greater than zero.
    c) equal to the battery's terminal voltage.
    d) zero.

    Answer: d Difficulty: 2

24. A resistor is connected to an AC power supply. On this circuit, the current
    a) leads the voltage by 90°.
    b) lags the voltage by 90°.
    c) is in phase with the voltage.
    d) none of the above.

    Answer: c Difficulty: 2

25. A pure inductor is connected to an AC power supply. In this circuit, the current
    a) leads the voltage by 90°.
    b) lags the voltage by 90°.
    c) is in phase with the voltage.
    d) none of the above.

    Answer: a Difficulty: 2

26. As the frequency of the AC voltage across a capacitor approaches zero, the capacitive reactance of that capacitor
    a) approaches zero.
    b) approaches infinity.
    c) approaches unity.
    d) none of the above.

    Answer: b Difficulty: 2

27. As the frequency of the AC voltage across an inductor approaches zero, the inductive reactance of that coil
    a) approaches zero.
    b) approaches infinity.
    c) approaches unity.
    d) none of the above.

    Answer: a Difficulty: 2

28. If the frequency of the AC voltage across a capacitor is doubled, the capacitive reactance of that capacitor
    a) increases to 4 times its original value.
    b) increases to twice its original value.
    c) decreases to one-half its original value.
    d) decreases to one-fourth its original value.

    Answer: c Difficulty: 2

29. If the frequency of the AC voltage across an inductor is doubled, the inductive reactance of that inductor
    a) increases to 4 times its original value.
    b) increases to twice its original value.
    c) decreases to one-half its original value.
    d) decreases to one-fourth its original value.

    Answer: b Difficulty: 2

TRUE/FALSE

30. There is no power loss in a pure LC circuit.

    Answer: T Difficulty: 2

31. A phase diagram is a graph of phase angle on the vertical axis and resistance on the horizontal axis.

    Answer: F Difficulty: 2

Chapter 20

MULTIPLE CHOICE

32. Consider an RLC circuit. The impedance of the circuit increases if R increases.
    a) Always true.
    b) True only if X_L is less than or equal to X_C.
    c) True only if X_L is greater than or equal to X_C.
    d) Never true.

    Answer: a Difficulty: 2

33. Consider an RLC circuit. The impedance of the circuit increases if X_L increases.
    a) Always true.
    b) True only if X_L is less than or equal to X_C.
    c) True only if X_L is greater than or equal to X_C.
    d) Never true.

    Answer: c Difficulty: 2

ESSAY

34. A series RC circuit has resistance of 150 θ and capacitance of 40 µF. It is driven by a 120-V, 60-Hz source. (a) How much power is dissipated in the resistor? (b) How much power is dissipated in the capacitor?

    Answer: (a) 88 W   (b) zero Difficulty: 2

35. A series RL circuit is driven by a 120-V, 60-Hz source. The values of the resistance and inductance are R = 20 Ω and L = 160 mH. Determine (a) the current. (b) phase angle between the current and the applied voltage.

    Answer: (a) I = 1.89 A   (b) I lags V by 72° Difficulty: 2

36. A 50-θ resistor is placed in series with a 40-mH inductor. At what frequency will the current in such a circuit lag the applied voltage by exactly 45°?

    Answer: 199 Hz Difficulty: 2

MULTIPLE CHOICE

37. The power factor of an RLC circuit is defined as $\cos \phi = R/Z$, but the phase angle can also be calculated from
    a) $\sin \phi = (X_L - X_C)/Z$
    b) $\sin \phi = (X_L - X_C)/RZ$
    c) $\sin \phi = Z/(X_L - X_C)$
    d) $\sin \phi = RZ/(X_L - X_C)$

    Answer: a Difficulty: 2

38. A series RLC circuit has R = 20 Ω, L = 200-mH, C = 10 μF. At what frequency should the circuit be driven in order to have maximum power transferred from the driving source?
    a) 113 Hz
    b) 167 Hz
    c) 277 Hz
    d) 960 Hz

    Answer: a Difficulty: 2

39. What size capacitor must be placed in series with a 30-θ resistor and a 40-mH coil if the resonant frequency of the circuit is to be 1000 Hz?
    a) 1.6 μ
    b) 2.0 μ
    c) 4.5 μ
    d) 6.0 μ

    Answer: a Difficulty: 2

40. Consider an RLC circuit that is driven by an AC applied voltage. At resonance,
    a) the peak voltage across the capacitor is greater than the peak voltage across the inductor.
    b) the peak voltage across the inductor is greater than the peak voltage across the capacitor.
    c) the current is in phase with the driving voltage.
    d) the peak voltage across the resistor is equal to the peak voltage across the inductor.

    Answer: c Difficulty: 2

41. Which of the following is a true statement?
    a) The voltage across a capacitor can change discontinuously.
    b) The current in an RC circuit can change instantaneously.
    c) The current in an RL circuit can change discontinuously.
    d) The energy stored in any circuit element can change instantaneously.

    Answer: b Difficulty: 2

Chapter 20

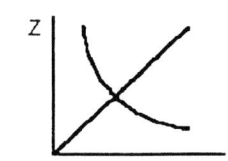

42. The graph, shown, represents resonance if the vertical axis is impedance and the
    a) horizontal axis is resistance.
    b) straight line is capacitive reactance, the curved line is inductive reactance, and the horizontal axis is frequency.
    c) straight line is inductive reactance, the curved line is capacitive reactance, and the horizontal axis is frequency.
    d) horizontal axis is inductance.

    Answer: c Difficulty: 2

43. In a given LC resonant circuit,
    a) the stored electric field energy is greater than the stored magnetic field energy.
    b) the stored electric field energy is less than the stored magnetic field energy.
    c) the stored electric field energy is equal to the stored magnetic field energy.
    d) all of the above are possible.

    Answer: d Difficulty: 2

44. Consider an RLC series circuit. The impedance of the circuit increases if X_C increases.
    a) Always true.
    b) True only if X_L is less than or equal to X_C.
    c) True only if X_L is greater than or equal to X_C.
    d) Never true.

    Answer: b Difficulty: 2

45. Resonance in a series RLC circuit occurs when
    a) X_L is greater than X_C.
    b) X_C is greater than X_L.
    c) $(X_L - X_C)^2$ is equal to R^2.
    d) Always true.

    Answer: b Difficulty: 2

46. A resistance of 55 Ω, a capacitor of capacitive reactance 30 Ω, and an inductor of inductive reactance 30 Ω are connected in series to a 110-V, 60-Hz power source. What current flows in this circuit?
    a) 2 A
    b) Less than 2 A
    c) More than 2 A
    d) None of the above

    Answer: a   Difficulty: 2

47. If the inductance and the capacitance both double in an LRC series circuit, the resonant frequency of that circuit will
    a) decrease to one-half its original value.
    b) decrease to one-fourth its original value.
    c) decrease to one-eighth its original value.
    d) none of the above.

    Answer: a   Difficulty: 2

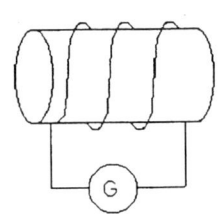

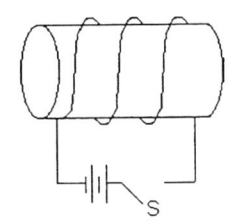

48. When the switch is closed, current is induced to flow through the Galvanometer
    a) toward the left
    b) toward the right

    Answer: a   Difficulty: 1

FILL-IN-THE-BLANK

49. To whom do we give credit for the 4 equations of electromagnetism? They are known as his equations (and contrary to popular opinion, he did not discover coffee).

    _____

    Answer: James Clerk MAXWELL   Difficulty: 1

Chapter 20

MULTIPLE CHOICE

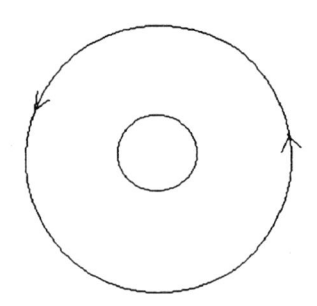

50. Consider an outer loop of wire surrounding an inner loop of wire. Suppose the current in the outer loop is counterclockwise and decreasing. What is the direction of the induced current in the inner loop?
    a) clockwise
    b) counterclockwise

    Answer: b Difficulty: 2

FILL-IN-THE-BLANK

51. An AC generator rotating at 60.Hertz in a 300.Gauss field has 1000 turns and produces a rms voltage of 150 volts and a rms current of 70.Amperes.

    a) What is the peak current?

    b) What is the area of each turn in the coil?

    Answer: a) 99.A   b) 0.019 m^2 Difficulty: 2

52. A transformer with 120.turns in its secondary supplies 12. volts at 220.ma to a toy train. The primary is connected across a 120.V wall outlet.

    a) How many turns are in the primary? _____

    b) What is the primary current? _____

    c) What power is delivered by the wall outlet? _____

    Answer: a) 1.2×10^3 b) 22.0 mA   c) 2.64 watts   Difficulty: 2

ESSAY

53. To transmit power over great distances, why is HIGH VOLTAGE chosen for long distance power lines?

    Answer:
    Power losses in the line go as $I^2 r$ (r being the wire resistance). By increasing the voltage, the current can be reduced, thereby greatly reducing losses.

    Difficulty: 2

MULTIPLE CHOICE

54. Visualize dropping a bar magnet through a horizontal loop of wire with the south magnetic pole entering the loop first. What is the direction of the induced current (viewed from above) as the north pole leaves the loop?
    a) clockwise
    b) counterclockwise
    c) zero

    Answer: b Difficulty: 2

FILL-IN-THE-BLANK

55. A magnetic field is generated inside a solenoid by a current in its coils. Suppose a change of temperature causes the diameter of the coils to increase by 1% but the length and current remain unchanged. What is the ratio of the new magnetic field in the solenoid to the original field?

    Answer: one (the n & I have not changed) Difficulty: 2

MULTIPLE CHOICE

56. Lenz's Law is a consequence of the law of conservation of:
    a) charge
    b) momentum
    c) electric field
    d) energy
    e) mass

    Answer: d Difficulty: 2

Chapter 20

ESSAY

57. Why won't a transformer work (stepup or stepdown potential) if a battery is connected across the input?

    Answer:
    It works by induction but a battery produces constant current and constant flux.
    Difficulty: 2

FILL-IN-THE-BLANK

58. Match the following wavelengths with the type electromagnetic wave:

_____	0.5 μm	A. Gamma Waves
_____	2.0 cm	B. Ultraviolet
_____	20. nm	C. Microwaves
_____	0.2 nm	D. Visible Light
_____	0.3 pm	E. X-rays
_____	3. m	F. FM radio

    Answer: D C B E A F   Difficulty: 2

MULTIPLE CHOICE

59. Whose law relates the rate of change of flux to an induced voltage?
    a) Lenz
    b) Volta
    c) Coulomb
    d) Faraday
    e) Hertz

    Answer: d Difficulty: 1

60. A loop of wire is rotated about a diameter (which is perpendicular to a given magnetic field.) In one revolution the induced current in the loop reverses direction how many times?
    a) 0
    b) 1
    c) 2
    d) 3
    e) 4
    f) none of the above

    Answer: c Difficulty: 2

ESSAY

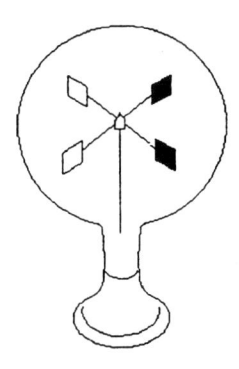

61. A popular novelty contains 4 vanes balanced on a needle in an evacuated globe (see diagram). The front side of each vane is a shiny mirror and the back side is blackened. Since electromagnetic waves exert a "radiation pressure", which way will the vanes rotate when a bright light shines on them?

Answer:
With a "good" vacuum, the vanes rotate ClockWise (from above) since light reflecting off the mirror transfers more momentum than the light absorbed on dark side. All too often the vacuum is poor so the vane rotates CounterClockWise: Radiation absorbed raises the temperature of the darkside and air molecules bouncing off gain energy and transfer momentum to the dark side.

Difficulty: 3

62. Why are transformer cores made of laminated, rather than solid, magnetic material?

Answer:
The changing flux in the transformer core tries to induce currents in the core material. These currents would lead to energy loss and undersirable heating effects. Lamination reduces the current flow and loss.

Difficulty: 2

FILL-IN-THE-BLANK

63. The starting current in a dc motor is 20.A, but it normally operates on 110.volts with a back EMF of 90.volts.
    A) What is the resistance in the windings?
    B) What is the current at normal load?

Answer: A) 5.5 ohms   B) 3.6 Amperes   Difficulty: 2

Chapter 20

64. A transformer has a turns ratio of 10. The primary voltage and current are 120.volts and 20.Amperes.
   A) What is the primary impedance (ratio of primary voltage to current)?
   B) What is the secondary impedance?

   Answer: A) 6.0 ohms   B) 0.60 KΩ   Difficulty: 2

ESSAY

65. Early AM radios required very long antennae wires. FM antennae are often approximately a meter in size. Can you surmise why sensitive FM aerials are much smaller than those AM wires?

   Answer:
   A sensitive aerial has a size related to the wave length of the radiation (e.g. λ, $\lambda/2$, $\lambda/4$). AM, having a frequency a thousand times lower than FM, has wavelengths a thousand times larger and early radios required a very sensitive antenna, unlike modern radios which have enormous amplification capability.

   Difficulty: 2

Chapter 21

MATCHING

Match the relationship to the physical phenomena.

a) $(\lambda_2/\lambda_1)(\sin\theta_1/\sin\theta_2) = 1$
b) $\arcsin(n_2/n_1)$
c) λ/λ_m
d) $\cong (d/n)$
e) $\theta_i = \theta_r$

1. Snell's law

   Answer: a) $(\lambda_2/\lambda_1)(\sin\theta_1/\sin\theta_2) = 1$ Difficulty: 2

2. critical angle

   Answer: b) $\arcsin(n_2/n_1)$ Difficulty: 2

3. index of refraction

   Answer: c) λ/λ_m Difficulty: 2

4. apparent depth

   Answer: d) $\cong (d/n)$ Difficulty: 2

5. Law of reflection

   Answer: e) $\theta_i = \theta_r$ Difficulty: 2

MULTIPLE CHOICE

6. Which of the following is a false statement?
   a) All points on a given wave front have the same phase.
   b) Rays are always perpendicular to wave fronts.
   c) All wave fronts have the same amplitude.
   d) The spacing between adjacent wave fronts is one-half wavelength.

   Answer: c   Difficulty: 2

ESSAY

7. A laser beam strikes a plane reflecting surface with an angle of incidence of 37°. What is the angle between the incident ray and the reflected ray?

   Answer: 74°   Difficulty: 2

Chapter 21

MULTIPLE CHOICE

8. A negative index of refraction for a medium would imply
   a) that the speed of light in the medium is the same as the speed of light in air.
   b) that the speed of light in the medium is greater than the speed of light in air.
   c) refraction is not possible.
   d) reflection is not possible.

   Answer: b Difficulty: 2

ESSAY

9. A light beam enters water at an angle of incidence of 37°. Determine the angle of refraction.

   Answer: 26.9° Difficulty: 2

10. Light of wavelength 550 nm in air is found to travel at 1.96×10^8 m/s in a certain liquid. Determine (a) the index of refraction of the liquid. (b) the frequency of the light in air. (c) the frequency of the light in the liquid. (d) the wavelength of the light in the liquid.

    Answer: (a) 1.53   (b) 5.45×10^{14} Hz   (c) 5.45×10^{14} Hz   (d) 359 nm

    Difficulty: 2

MULTIPLE CHOICE

11. When a light wave enters into a medium of different optical density,
    a) its speed and frequency change.
    b) its speed and wavelength change.
    c) its frequency and wavelength change.
    d) its speed, frequency, and wavelength change.

    Answer: b Difficulty: 2

TRUE/FALSE

12. Optical density is directly proportional to mass density.

    Answer: F Difficulty: 2

13. Laser light is invisible in air.

    Answer: T Difficulty: 2

14. When light shines on the surface of this page, the reflection is diffuse.

Answer: T Difficulty: 2

ESSAY

15. A buoy, used to mark a harbor channel, consists of a weighted rod 2 m in length. 1.5 m of the rod is immersed in water and 0.5 m extends above the surface. If sunlight is incident on the water with an angle of incidence of 40°, what is the length of the shadow of the buoy on the level bottom of the harbor?

Answer: 1.25 m Difficulty: 2

16. A light ray strikes a glass plate of thickness 0.8 cm at an angle of incidence of 60°. The index of refraction of the glass is 1.55. By how much is the beam displaced from its original line of travel after it has passed through the glass?

Answer: 0.42 cm Difficulty: 2

MULTIPLE CHOICE

17. Light travels fastest
    a) in a vacuum.
    b) through water.
    c) through glass.
    d) through diamond.

Answer: a Difficulty: 2

18. When a beam of light (wavelength = 590 nm), originally traveling in air, enters a piece of glass (index of refraction 1.50), its frequency
    a) increases by a factor of 1.50.
    b) is reduced to 2/3 its original value.
    c) is unaffected.
    d) none of the above.

Answer: c Difficulty: 2

19. When a beam of light (wavelength = 590 nm), originally traveling in air, enters a piece of glass (index of refraction 1.50), its wavelength
    a) increases by a factor of 1.50.
    b) is reduced to 2/3 its original value.
    c) is unaffected.
    d) none of the above.

Answer: b Difficulty: 2

Chapter 21

20. The index of refraction of diamond is 2.42. This means that a given wavelength of light travels
    a) 2.42 times faster in air than it does in diamond.
    b) 2.42 times faster in diamond than it does in air.
    c) 2.42 times faster in vacuum than it does in diamond.
    d) 2.42 times faster in diamond than it does in vacuum.

    Answer: c Difficulty: 2

21. A beam of light (f = 5 x 10^{14} Hz) enters a piece of glass (n = 1.5). What is the frequency of the light while it is in the glass?
    a) 5 x 10^{14} Hz
    b) 7.5 x 10^{14} Hz
    c) 3.33 x 10^{14} Hz
    d) None of the above

    Answer: a Difficulty: 2

22. A ray of light, which is traveling in air, is incident on a glass plate at a 45° angle. The angle of refraction in the glass
    a) is less than 45°.
    b) is greater than 45°.
    c) is equal to 45°.
    d) could be any of the above; it all depends on the index of refraction of glass.

    Answer: a Difficulty: 2

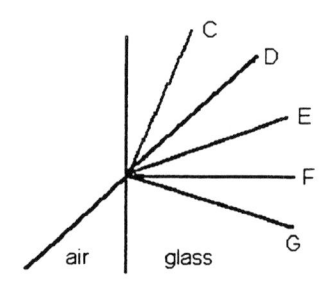

23. Shown here are some possible paths that a light ray can follow in going either from glass to air or from air into glass. In the drawing you are not told whether the light is going from left to right or from right to left. Which path did the light follow?
    a) One cannot answer this question without knowing in which direction the light is going.
    b) More than one of these paths is possible.
    c) Path C
    d) Path D
    e) Path E
    f) Path F
    g) Path G

    Answer: e Difficulty: 2

24. A ray of light, which is traveling in a vacuum, is incident on a glass plate (n = ng). For increasing angles of incidence, the angle of refraction
    a) increases, approaching the limiting value of 90°.
    b) increases, approaching the limiting value of $\sin^{-1}(1/n_g)$ degrees.
    c) increases, approaching the limiting value of $\sin^{-1}(n_g)$ degrees.
    d) decreases, approaching the limiting value of zero degrees.

    Answer: b Difficulty: 2

25. When a ray of light passes obliquely from one medium to another, which of the following changes?
    a) Direction of travel
    b) Wavelength
    c) Speed
    d) All of the above

    Answer: d Difficulty: 2

26. For all transparent material substances, the index of refraction
    a) is less than 1.
    b) is greater than 1.
    c) is equal to 1.
    d) could be any of the above; it all depends on optical density.

    Answer: b Difficulty: 2

27. The angle of incidence
    a) must equal the angle of refraction.
    b) is always less than the angle of refraction.
    c) is always greater than the angle of refraction.
    d) may be greater than, less than, or equal to the angle of refraction.

    Answer: d Difficulty: 2

28. The angle of incidence
    a) must equal the angle of reflection.
    b) is always less than the angle of reflection.
    c) is always greater than the angle of reflection.
    d) may be greater than, less than, or equal to the angle of reflection.

    Answer: a Difficulty: 2

29. Which color of light undergoes the greatest refraction when passing from air to glass?
    a) Red
    b) Yellow
    c) Green
    d) Blue

    Answer: d Difficulty: 2

Chapter 21

TRUE/FALSE

30. White light is monochromatic light.

   Answer: F Difficulty: 2

MULTIPLE CHOICE

31. A beam of white light is incident on a thick glass plate with parallel sides, at an angle between 0° and 90° with the normal. Which color emerges from the other side first?
   a) Red
   b) Green
   c) Violet
   d) None of the above. All colors emerge at the same time.

   Answer: a Difficulty: 2

TRUE/FALSE

32. When white light enters a more optically dense medium than that from which it came, the green color component refracts more than the yellow color component.

   Answer: T Difficulty: 2

MULTIPLE CHOICE

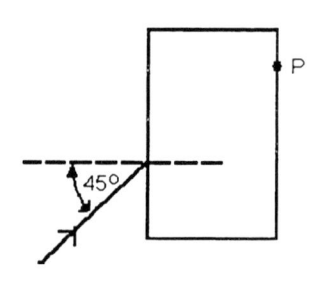

33. A light beam composed of red and blue light is incident upon a rectangular glass plate, as shown. The light emerges into the air from point P as two separate beams
   a) that are parallel, with the red beam displaced below the blue beam.
   b) that are parallel, with the blue beam displaced below the red beam.
   c) that are not parallel, with the blue beam displaced below the red beam.
   d) that are not parallel, with the red beam displaced below the blue beam.
   e) none of the above is correct.

   Answer: b Difficulty: 2

34. An oil layer that is 5 cm thick is spread smoothly and evenly over the surface of water on a windless day. What is the angle of refraction in the water for a ray of light that has an angle of incidence of 45° as it enters the oil from the air above? (The index of refraction for oil is 1.15, and for water it is 1.33.)
   a) 27.2°
   b) 32.1°
   c) 35.5°
   d) 38.6°

   Answer: b Difficulty: 2

35. A beam of light, traveling in air, strikes a plate of transparent material at an angle of incidence of 56°. It is observed that the reflected and refracted beams form an angle of 90°. What is the index of refraction of this material?
   a) n = 1.40
   b) n = 1.43
   c) n = 1.44
   d) n = 1.48

   Answer: d Difficulty: 2

ESSAY

36. A point source of light is positioned 20 m below the surface of a lake. What is the diameter of the largest circle on the surface of the water through which light can emerge?

   Answer: 45.6 m Difficulty: 2

37. The end of a cylindrical plastic rod is polished and cut perpendicular to the axis of the cylinder. Determine the minimum index of refraction so that a light ray entering the end of the rod will always be totally internally reflected within the rod, i.e., it will never escape the rod until it comes to the other end.

   Answer: 1.414 Difficulty: 2

MULTIPLE CHOICE

38. The critical angle for a beam of light passing from water into air is 48.8°. This means that all light rays with an angle of incidence greater than this angle will be
   a) absorbed.
   b) totally reflected.
   c) partially reflected and partially transmitted.
   d) totally transmitted.

   Answer: b Difficulty: 2

Chapter 21

39. What is the critical angle for light traveling from crown glass (n = 1.52) into water (n = 1.33)?
   a) 42°
   b) 48°
   c) 57°
   d) 61°

   Answer: d Difficulty: 2

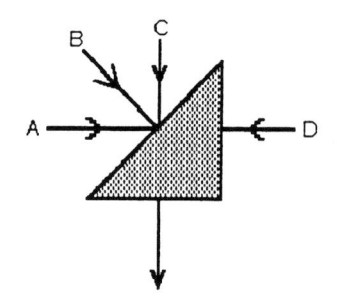

40. A light ray emerges from the bottom of a 45° glass prism into air, as shown in the sketch. From which direction was the ray incident on the prism?
   a) A
   b) B
   c) C
   d) D

   Answer: d Difficulty: 2

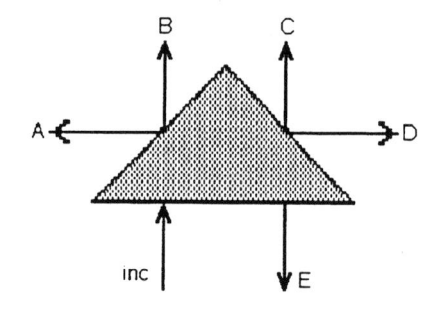

41. A ray of light (inc) enters a 45° - 90° glass prism from air (bottom left), as shown. In what direction does the light re-emerge back into the air?
   a) A
   b) B
   c) C
   d) D
   e) E

   Answer: e Difficulty: 2

42. An optical fiber is 1 meter long and has a diameter of 20 μm. Its ends are perpendicular to its axis. Its index of refraction is 1.30. What is the minimum number of reflections a light ray entering one end will make before it emerges from the other end?
    a) 28,500
    b) 25,600
    c) 24,220
    d) 18,500

    Answer: d Difficulty: 2

43. Fiberscopes are based on the physical phenomena called
    a) absorption.
    b) dispersion.
    c) total internal reflection.
    d) diffusion.

    Answer: c Difficulty: 2

44. White light is
    a) light of wavelength 550 nm, in the middle of the visible spectrum.
    b) a mixture of all frequencies.
    c) a mixture of red, green, and blue light.
    d) the term used to describe very bright light.
    e) the opposite (or complementary color) of black light.

    Answer: b Difficulty: 2

45. The person who proved that white light is composed of light of different colors is
    a) Willebrod Snell.
    b) Christen Huygens.
    c) Isaac Newton.
    d) Thomas Edison.
    e) General Electric.

    Answer: c Difficulty: 2

ESSAY

46. A parallel light beam containing two wavelengths, 480 nm and 700 nm, strikes a plain piece of glass at an angle of incidence of 60°. The index of refraction of the glass is 1.4830 at 480 nm and 1.4760 at 700 nm. Determine the angle between the two beams in the glass.

    Answer: 0.196° Difficulty: 2

Chapter 21

MULTIPLE CHOICE

47. A ray of light consisting of blue light (wavelength 480 nm) and red light (wavelength 670 nm) is incident on a thick piece of glass at 80°. What is the angular separation between the refracted red and refracted blue beams while they are in the glass? (The respective indices of refraction for the blue light and the red light are 1.4636 and 1.4561.)
    a) 0.277°
    b) 0.330°
    c) 0.341°
    d) 0.455°

    Answer: a   Difficulty: 2

48. White light, coming from medium 1 (index of refraction n_1), will disperse upon entering medium 2 (index of refraction n_2) when
    a) $(n_2 > n_1)$
    b) $(n_2 < n_1)$
    c) $(n_2 = n_1)$
    d) $(n_2 > n_1)$ or $(n_2 < n_1)$

    Answer: d   Difficulty: 2

TRUE/FALSE

49. You can only see a rainbow if the sun is behind you.

    Answer: T   Difficulty: 2

50. The higher the sun is up in the sky, the more the rainbow can be seen from the ground.

    Answer: F   Difficulty: 2

51. The color on the outer edge of the primary rainbow is red.

    Answer: T   Difficulty: 2

52. The color on the outer edge of a rainbow is the same for both primary and secondary rainbows.

    Answer: F   Difficulty: 2

FILL-IN-THE-BLANK

53. The current in a coil increases from 2.0 A to 12.A in 50. ms. Calculate the voltage induced in the coil if its self-inductance is 0.120 H.

    Answer: 24 volts opposing the current   Difficulty: 2

54. At what frequency will the Inductive Reactance of 44.mH be equal to the Capacitive Reactance of a 27.pF capacitor?

Answer: 0.15 MHz Difficulty: 2

MULTIPLE CHOICE

55. Consider a 4500.ohm resistor in series with a 133.mH inductor. What capacitance should be added is series so that the impedance is a minimum at 500.KHz ?
    a) 0.267 pF
    b) 0.627 pF
    c) 0.762 pF
    d) 2.67 µF
    e) 7.26 µF

Answer: c Difficulty: 2

56. Consider a series circuit in which the capacitive reactance equals the inductive reactance. What is the phase angle between current and voltage?
    a) -90°
    b) +90°
    c) -180°
    d) +180°
    e) 0°

Answer: e Difficulty: 2

ESSAY

57. Choose 2 (or more) different numbers and show that their average is different from the rms value.

Answer: For example consider 3 and 4: avg=3.50 , rms=3.54 Difficulty: 2

TRUE/FALSE

58. The impedance of an ac circuit is 50.ohms and its rms current is 2.0 amperes so the power dissipated is 200.watts.

Answer: T Difficulty: 1

Chapter 21

MULTIPLE CHOICE

59. An AC circuit has a 100.ohm resistor in series with a 4.9 μF capacitor and a 700.mH inductor. At what frequency does the circuit act like a pure resistance?
    a) 86.Hz
    b) 540.Hz
    c) 1.9 mHz
    d) 12.mHz
    e) 0.29 MHz

    Answer: a Difficulty: 2

60. Consider a capacitor connected across the AC supply. The current through the capacitor will _____ as the capacitor is increased.
    a) increase
    b) remain constant
    c) decrease

    Answer: a Difficulty: 1

FILL-IN-THE-BLANK

61. American power plants usually supply 120.volts AC:
    a) at what frequency?
    b) at what maximum voltage?

    Answer: a) 60.Hz   b) 170.volts if rms=120v Difficulty: 2

62. A 90.% efficient transformer supplies 3.0 Amperes at 6.0 volts. If the input voltage is 120.v, what is the input current?

    Answer: 0.17 A Difficulty: 2

TRUE/FALSE

63. The average power dissipated in an RLC circuit is $I^2_{rms}R$ where R is the Reactance.

    Answer: F Difficulty: 1

Testbank

MULTIPLE CHOICE

64. The unit of Inductive Reactance is:
    a) Henry
    b) Ohm-Henry
    c) Ohm
    d) Ohm-meter
    e) Henry-meter
    f) Henry/meter

    Answer: c   Difficulty: 1

65. A resistor is connected to normal 60.Hz - 120.volt AC. How many times a second does the instantaneous dissipated POWER reach a maximum?
    a) 0
    b) 30
    c) 60
    d) 120
    e) 240

    Answer: d   Difficulty: 2

66. In an RLC circuit, the resistance is 10.ohms, the capacitive reactance is 30.ohms, and the inductive reactance is 50.ohms. What is the circuit impedance?
    a) 10.Ω
    b) 22.Ω
    c) 63.Ω
    d) 81.Ω
    e) 90.Ω

    Answer: b   Difficulty: 2

FILL-IN-THE-BLANK

67. In an AC circuit the voltage varies as V=96.sin(277 t)
    a) What is the rms voltage ?
    b) What is the frequency ?

    Answer: a) 68.v   b) 44.1 Hz   Difficulty: 2

Chapter 21

MULTIPLE CHOICE

68. If I have a circuit which has unwanted current variations (with time) and I want to "smooth them out" (want a more steady current), which one of these should be added in series?
a) capacitor
b) resistance
c) inductor
d) battery

Answer: c Difficulty: 2

69. The average power delivered by an AC generator in an RLC circuit is _?_ the product IV where I & V are respectively the current and voltage amplitudes.
a) more than
b) equal to
c) less than

Answer: c Difficulty: 2

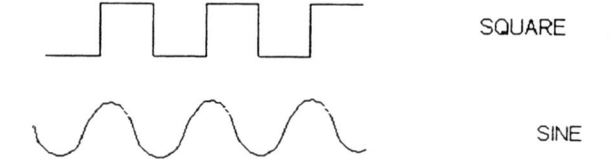

SQUARE

SINE

70. Is the rms value of a "square" wave _____ the rms value of a "sine" wave of similar amplitude and frequency? (see figure)
a) more than
b) equal to
c) less than

Answer: a Difficulty: 2

Chapter 22

MULTIPLE CHOICE

1. A plane mirror forms an image that is
   a) real and upright.
   b) virtual and upright.
   c) real and upside down.
   d) virtual and upside down.

   Answer: b Difficulty: 2

TRUE/FALSE

2. If you place a plane mirror vertically into water, and a dolphin moves toward it with a speed of 4 m/s, she will see her image move toward her at 4 m/s.

   Answer: F Difficulty: 2

MULTIPLE CHOICE

   a) AMBULANCE (inverted)   b) AMBULANCE (mirrored)

   c) ECNALUBMA   d) AMBULANCE (rotated)

   e) AMBULANCE (upside-down mirrored)

3. You may have seen ambulances on the street with the letters of the word AMBULANCE written on the front of them, in such a way as to appear correctly when viewed in your car's rear-view mirror. How do the letters appear when you look directly at the ambulance?

   Answer: e Difficulty: 2

4. The simplest type of camera is a pin-hole camera; you might have even made one. It consists of a box with a single pinhole in it, and a piece of film on the inner side opposite the hole. Suppose you photograph your friend with it. Your friend's image on the film will appear
   a) upside-down because light travels in a straight line.
   b) right-side up because light travels in a straight line.
   c) upside-down because light refracts at the pinhole.
   d) right-side up because light refracts at the pinhole.

   Answer: a Difficulty: 2

Chapter 22

ESSAY

5. A man 190 cm tall stands 25 m from a plane wall mirror. One meter in front of him sits his dog, 30 cm tall. What minimum mirror size is needed for the man to see all of his dog's reflection?

   Answer: 18.8 cm tall   Difficulty: 2

6. A friend who is 180 cm tall stands 2 m from a plane wall mirror. You stand slightly behind him, 3 m from the wall mirror. How tall must the mirror be if you are to see all of your friend in the mirror?

   Answer: 1.08 m   Difficulty: 2

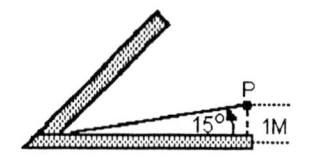

7. At an amusement park you stand between two lane mirrors, which intersect at an angle of 60°. You are positioned at point P in the drawing here. You are 1 m from the nearest mirror, and 15° above it. What is the distance from you to the nearest image and to the second nearest image?

   Answer: 2 m; 5.5 m   Difficulty: 2

8. Two plane mirrors intersect at an angle of 120°. Draw ray diagrams to determine the locations of all the images formed by an object placed at a point P between the mirrors, closer to one mirror than the other.

   Answer:

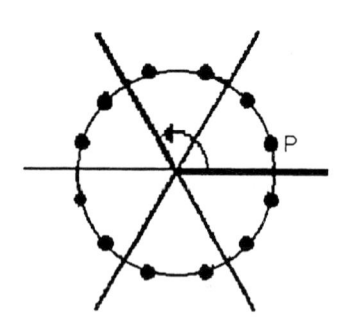

   Difficulty: 2

MULTIPLE CHOICE

9. I recently had to replace the side view mirror on my truck. How tall should the mirror have been if I am to see all of a car 2 m high when it is following 20 m back (measured from the mirror). The mirror is 0.5 m from my eyes.
   a) 4.9 cm
   b) 5.2 cm
   c) 5.8 cm
   d) 7.0 cm

   Answer: a Difficulty: 2

ESSAY

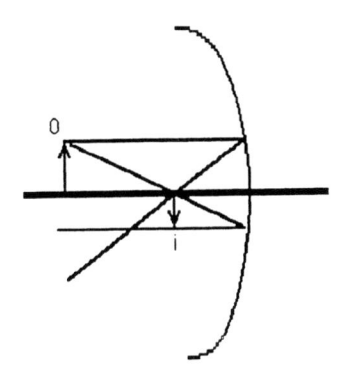

10. An object is placed 6 cm from a concave mirror of radius 4 cm. Graphically determine (a) the image position. (b) if the image is larger, smaller, or the same size as the object. (c) if the image is real or virtual.

    Answer:
    (a) (shown)
    (b) smaller
    (c) real

    Difficulty: 2

Chapter 22

MULTIPLE CHOICE

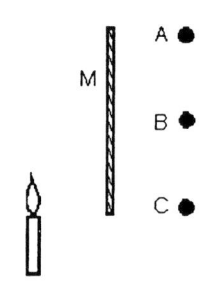

11. Suppose a lighted candle is placed a short distance from a plane mirror, as shown here. Where will the image of the flame be located?
    a) At A
    b) At B
    c) At C
    d) At M (at the mirror)

    Answer: c Difficulty: 2

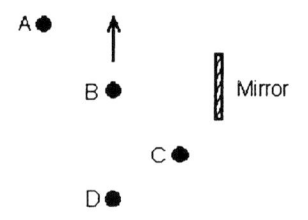

12. At which, if any, of the points indicated here should you place your eye if you wish to see an image of the arrow in the mirror?
    a) A
    b) B
    c) C
    d) D
    e) None of the above

    Answer: d Difficulty: 2

13. An object is placed a distance d in front of a plane mirror. The size of the image will be
    a) half as big as the size of the object.
    b) dependent on the distance d.
    c) dependent on where you are positioned when you look at the image.
    d) twice the size of the object.
    e) the same size as the object, independent of the distance d or the position of the observer.

    Answer: e Difficulty: 2

TRUE/FALSE

14. The rear-view mirrors on the passenger side of many new cars have a warning written on them: "OBJECTS IN MIRROR ARE CLOSER THAN THEY APPEAR." This implies that the mirror must be concave.

    Answer: T   Difficulty: 2

MULTIPLE CHOICE

15. If you stand in front of a concave mirror, exactly at its focal point,
    a) you won't see your image because there is none.
    b) you won't see your image because it's focused at a different distance.
    c) you will see your image, and you will appear smaller.
    d) you will see your image and you will appear larger.
    e) you will see your image at your same height.

    Answer: a   Difficulty: 2

16. If you stand in front of a concave mirror, exactly at its center of curvature,
    a) you won't see your image because there is none.
    b) you won't see your image because it's focused at a different distance.
    c) you will see your image and you will appear smaller.
    d) you will see your image and you will appear larger.
    e) you will see your image at your same height.

    Answer: e   Difficulty: 2

17. If you stand in front of a convex mirror, at the same distance from it as its focal length,
    a) you won't see your image because there is none.
    b) you won't see your image because it's focused at a different distance.
    c) you will see your image and you will appear smaller.
    d) you will see your image and you will appear larger.
    e) you will see your image at your same height.

    Answer: a   Difficulty: 2

18. If you stand in front of a convex mirror, at the same distance from it as its radius of curvature,
    a) you won't see your image because there is none.
    b) you won't see your image because it's focused at a different distance.
    c) you will see your image and you will appear smaller.
    d) you will see your image and you will appear larger.
    e) you will see your image at your same height.

    Answer: c   Difficulty: 2

Chapter 22

ESSAY

19. An object 2 cm tall is placed 24 cm in front of a convex mirror whose focal length is 30 cm. (a) Where is the image formed? (b) How tall is it?

    Answer: (a) 14.4 cm behind the mirror   (b) 1.2 cm tall   Difficulty: 2

20. An object is placed 15 cm from a concave mirror of focal length 20 cm. If the object is 4 cm tall, (a) where is the image formed? (b) how tall is it?

    Answer: (a) 60 cm behind the mirror   (b) 16 cm tall   Difficulty: 2

21. In using ray tracing to graphically locate the image of an object that is placed in front of a mirror, describe three simple rays that you could draw, that pass by the head of the object.

    Answer:
    Ray 1: the radial ray; i.e., the ray drawn toward the center of curvature, which is then reflected back along the same path.
    Ray 2: the parallel ray; i.e., the incoming ray parallel to the main axis, which is then reflected along a path that extends to the focal point.
    Ray 3: the focal ray; i.e., the ray drawn toward the focal point, which is then reflected back parallel to the main axis.

    Difficulty: 2

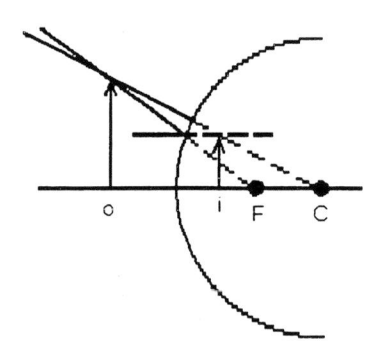

22. An object is placed 3 cm from a convex mirror of radius 4 cm. (a) Graphically determine the size and position of the image. (b) Is the image real or virtual?

    Answer:
    (a) (see sketch)
    (b) virtual
    Difficulty: 2

MULTIPLE CHOICE

23. Sometimes when you look into a curved mirror you see a magnified image (a great big you!) and sometimes you see a diminished image (a little you). If you look at the bottom (convex) side of a shiny spoon, what will you see?
    a) You won't see an image of yourself because no image will be formed.
    b) You will see a little you, upside down.
    c) You will see a little you, right side up.
    d) You will see a little you, but whether you are right side up or upside down depends on how near you are to the spoon.
    e) You will either see a little you or a great big you, depending on how near you are to the spoon.

    Answer: c Difficulty: 2

24. Which of the following is an accurate statement?
    a) A mirror always forms a real image.
    b) A mirror always forms a virtual image.
    c) A mirror always forms an image larger than the object.
    d) A mirror always forms an image smaller than the object.
    e) None of the above is true.

    Answer: e Difficulty: 2

25. A negative magnification for a mirror means
    a) the image is inverted, and the mirror is concave.
    b) the image is inverted, and the mirror is convex.
    c) the image is inverted, and the mirror may be concave or convex.
    d) the image is upright, and the mirror is convex.
    e) the image is upright, and the mirror may be concave or convex.

    Answer: c Difficulty: 2

ESSAY

26. In using ray tracing to graphically locate the image of an object that is placed in front of a lens, describe three simple rays that you could draw, that pass by the head of the object.

    Answer:
    Ray 1: the vertex ray; i.e., the ray passing through the center of the lens, which emerges undeflected.
    Ray 2: the parallel ray; i.e., the incoming ray parallel to the main axis, which is then refracted along a path that extends to the focal point.
    Ray 3: the focal ray; i.e., the ray drawn toward the focal point, which is then refracted parallel to the main axis.
    Difficulty: 2

Chapter 22

27. An object is placed 9.5 cm from a lens of focal length 24 cm. (a) Where is the image formed? (b) What is the magnification?

   Answer: (a) 16 cm from the lens on the same side as the object. (b) m = 1.7

   Difficulty: 2

MULTIPLE CHOICE

28. The image of the rare stamp you see through a magnifying glass is
    a) always the same orientation as the stamp.
    b) always upside-down compared to the stamp.
    c) either the same orientation or upside-down, depending on how close the stamp is to the glass.
    d) either the same orientation or upside-down, depending on the thickness of the glass used.

    Answer: c Difficulty: 2

ESSAY

29. Where must an object be placed with respect to a converging lens of focal length 30 cm if the image is to be virtual, and three times as large as the object?

    Answer: 20 cm from the lens Difficulty: 2

30. A camera with a telephoto lens of focal length 125 mm is used to take a photograph of a plant 1.8 m tall. The plant is 5.0 m from the lens.
    (a) What must be the distance between the lens and the camera film if the image is sharply focused? (b) How tall is the image?

    Answer: (a) 128 mm from lens to film   (b) 45 mm tall Difficulty: 2

TRUE/FALSE

31. A diverging lens is thicker at its center than it is at its edges.

    Answer: F Difficulty: 2

MULTIPLE CHOICE

32. The images formed by concave lenses
    a) are always real.
    b) are always virtual.
    c) could be real or virtual; it depends on whether the object distance is smaller or greater than the focal length.
    d) could be real or virtual, but always real when the object is placed at the focal point.

    Answer: b Difficulty: 2

ESSAY

33. A slide projector has a lens of focal length 15 cm. An image 1 m x 1 m is formed of a slide whose dimensions are 5 cm x 5 cm. How far from the lens must the screen be placed?

    Answer: 3.15 m Difficulty: 2

MULTIPLE CHOICE

34. How far from a 50-mm focal length lens, such as is used in many 35-mm cameras, must an object be positioned if it is to form a real image magnified in size by a factor of three?
    a) 46.2 mm
    b) 52.2 mm
    c) 57.5 mm
    d) 66.7 mm

    Answer: d Difficulty: 2

35. How far from a lens of focal length 50 mm must the object be placed if it is to form a virtual image magnified in size by a factor of three?
    a) 33.3 mm
    b) 42.2 mm
    c) 48.0 mm
    d) 54.4 mm

    Answer: a Difficulty: 2

TRUE/FALSE

36. If you stand in front of a double-concave lens, exactly at its center of curvature, your image will appear where you are, but upside-down.

    Answer: T Difficulty: 2

Chapter 22

ESSAY

37. An object is placed at the origin. A converging lens of focal length 10 mm is placed at x = 40 mm on the x axis. A second converging lens of focal length 20 mm is placed at x = 90 mm. Graphically determine the size and location of the final image.

    Answer:

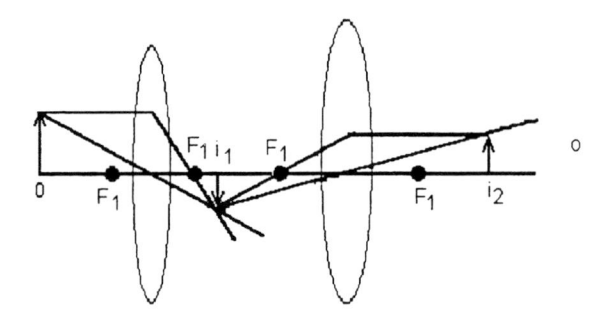

    Difficulty: 2

MULTIPLE CHOICE

38. A lamp is placed 1 m from a screen. Between the lamp and the screen is placed a converging lens of focal length 24 cm. The filament of the lamp can be imaged on the screen. As the lens position is varied with respect to the lamp,
    a) no sharp image will be seen for any lens position.
    b) a sharp image will be seen when the lens is halfway between the lamp and the screen.
    c) a sharp image will be seen when the lens is 40 cm from the lamp.
    d) a sharp image will be seen when the lens is 60 cm from the lamp.
    e) a sharp image will be seen when the lens is either 40 cm from the lamp or 60 cm from the lamp, but not otherwise.

    Answer: e Difficulty: 2

39. A diverging lens (f = -4 cm) is positioned 2 cm to the left of a converging lens (f = +6 cm). A 1-mm diameter beam of parallel light rays is incident on the diverging lens from the left. After leaving the converging lens, the outgoing rays
    a) converge.
    b) diverge.
    c) form a parallel beam of diameter D > 1 mm.
    d) form a parallel beam of diameter D < 1 mm.
    e) will travel back toward the light source.

    Answer: c Difficulty: 2

40. Is it possible to see a virtual image?
    a) No, since the rays that seem to emanate from a virtual image do not in fact emanate from the image.
    b) No, since virtual images do not really exist.
    c) Yes, the rays that appear to emanate from a virtual image can be focused on the retina just like those from an illuminated object.
    d) Yes, since almost everything we see is virtual because most things do not themselves give off light, but only reflect light coming from some other source.
    e) Yes, but only indirectly in the sense that if the virtual image is formed on a sheet of photographic film, one could later look at the picture formed.

    Answer: c Difficulty: 2

41. Two thin double-convex (convex-convex) lenses are placed in contact. If each has a focal length of 20 cm, how would you expect the combination to function?
    a) About like a single lens of focal length 20 cm
    b) About like a single lens of focal length 40 cm
    c) About like a single lens of focal length slightly greater than 20 cm
    d) About like a single lens of focal length less than 20 cm

    Answer: d Difficulty: 2

42. In a single-lens reflex camera the lens-film distance may be varied by sliding the lens forward or backward with respect to the camera housing. If, with such a camera, a fuzzy picture is obtained, this means that
    a) the lens was too far from the film.
    b) the lens was too close to the film.
    c) too much light was incident on the film.
    d) too little light was incident on the film.
    e) one cannot say which of the above reasons is valid.

    Answer: e Difficulty: 2

43. Two very thin lenses, each with focal length 20 cm, are placed in contact. What is the focal length of this compound lens?
    a) 40 cm
    b) 20 cm
    c) 15 cm
    d) 10 cm

    Answer: d Difficulty: 2

44. Two thin lenses, of focal lengths f_1 and f_2 placed in contact with each other are equivalent to a single lens of focal length
    a) $f_1 + f_2$
    b) $1/(f_1 + f_2)$
    c) $(f_1 + f_2)/f_1 f_2$
    d) $f_1 f_2/(f_1 + f_2)$

    Answer: d Difficulty: 2

Chapter 22

45. When an object is placed 60 cm from a converging lens, it forms a real image. When the object is moved to 40 cm from the lens, the image moves 10 cm farther from the lens. What is the focal length of the lens?
a) 20 cm
b) 30 cm
c) 40 cm
d) 42 cm

Answer: d Difficulty: 2

ESSAY

46. A converging lens with the same curvature on both sides and focal length 25 cm is to be made from crown glass (n = 1.52). What radius of curvature is required for each face?

Answer: 26 cm Difficulty: 2

47. A double convex lens has faces of radii 18 cm and 20 cm. When an object is placed 24 cm from the lens, a real image is formed 32 cm from the lens. Determine (a) the focal length of the lens (b) the index of refraction of the lens material.

Answer: (a) 13.7 cm   (b) n = 1.69 Difficulty: 2

MULTIPLE CHOICE

48. A double convex (convex-convex) thin lens has radii of curvature 46 cm, and is made of glass of index of refraction n = 1.60. What is the focal length?
a) infinite
b) 36 cm
c) 30 cm
d) 18 cm

Answer: c Difficulty: 2

49. A plano-convex lens is to have a focal length of 40 cm. It is made of glass of index of refraction 1.65. What radius of curvature is required?
a) 13 cm
b) 26 cm
c) 32 cm
d) 36 cm

Answer: b Difficulty: 2

50. Spherical lenses suffer from
   a) both spherical and chromatic aberration.
   b) spherical aberration, but not chromatic aberration.
   c) chromatic aberration, but not spherical aberration.
   d) neither spherical nor chromatic aberration.

   Answer: a Difficulty: 2

TRUE/FALSE

51. Spherical aberration can occur in mirrors as well as lenses.

   Answer: T Difficulty: 2

MULTIPLE CHOICE

52. Spherical mirrors suffer from
   a) both spherical and chromatic aberration.
   b) spherical aberration, but not chromatic aberration.
   c) chromatic aberration, but not spherical aberration.
   d) neither spherical nor chromatic aberration.

   Answer: b Difficulty: 2

TRUE/FALSE

53. Chromatic aberration occurs because of dispersion.

   Answer: T Difficulty: 2

MULTIPLE CHOICE

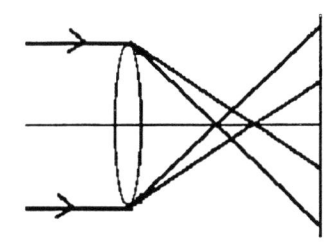

54. When two parallel white rays pass through the outer edges of a converging glass lens, chromatic aberration will cause colors to appear on the screen in the following order, from the top down:
   a) red, blue, red, blue.
   b) red, blue, blue, red.
   c) blue, red, blue, red.
   d) blue, red, red, blue.
   e) blue, blue, red, red.

   Answer: d Difficulty: 2

Chapter 22

55. Compare two diverging lenses similar except that lens B is rated at 20 diopters, whereas lens A is rated at 10 diopters. The focal length of lens B is
   a) one-fourth of the focal length of lens A.
   b) one-half of the focal length of lens A.
   c) twice the focal length of lens A.
   d) four times the focal length of lens A.

   Answer: b Difficulty: 2

TRUE/FALSE

56. A +20 diopter lens has a different power in water.

   Answer: T Difficulty: 2

MULTIPLE CHOICE

57. What is the index of refraction inside a certain transparent material if the critical angle is found to be 48.degrees (air is outside)?

   a) 1.22
   b) 1.35
   c) 1.48
   d) 1.49
   e) 0.743
   f) 3×10^8

   Answer: b Difficulty: 2

58. When a light beam enters an "optically more dense" material, it is bent:
   a) toward the normal to the surface
   b) away from the normal to the surface
   c) along the normal to the surface

   Answer: a Difficulty: 2

59. The IMAGE of a plane mirror (of a real object) has the following characteristics:
   a) real, erect, and magnification = 1
   b) virtual, erect, and magnification = 1
   c) real, inverted, and magnification < 1
   d) virtual, inverted, and magnification > 1

   Answer: b Difficulty: 2

60. For certain angles, TOTAL INTERNAL REFLECTION can occur when light tries to pass from one medium to another which has:
   a) a larger index of refraction
   b) a smaller index of refraction
   c) the same index of refraction
   d) an index of refraction which is $\sin^{-1}(45°)$

   Answer: b Difficulty: 1

FILL-IN-THE-BLANK

61. If a transparent material has an index of refraction of 1.67, what is the speed of light in the material?

   Answer: $0.599c = 1.80 \times 10^8$ m/s Difficulty: 1

MULTIPLE CHOICE

62. Which of the following colors undergoes the greatest REFRACTION when passing from air into glass? (note the accompanying dispersion curve)
   a) yellow
   b) red
   c) blue
   d) green
   e) yellow

   Answer: c Difficulty: 1

TRUE/FALSE

63. When TOTAL REFLECTION occurs at a surface, the incident angle must be in the substance with the higher speed (of light).

   Answer: F Difficulty: 2

MULTIPLE CHOICE

64. When the index of refraction of a transparent material varies with wavelength, the material exhibits:
   a) polarization
   b) reflection
   c) dispersion
   d) total internal reflection
   e) density variation

   Answer: c Difficulty: 1

Chapter 22

FILL-IN-THE-BLANK

65. The critical angle is given by $\sin_c(1/n_1)$ for what condition?

    Answer: The second index $(n_2) = 1$, equivalent to a vacuum. Difficulty: 1

ESSAY

66. The Apollo missions left mirrors on the moon so laser beams could be reflected off the moon from the Earth. The moon wobbles so a regular mirror would seldom be lined up correctly, yet the mirrors are always correctly aligned. How do you suppose NASA accomplished this?

    Answer:
    They used "corner reflectors" shaped like the vertex of a cube. A ray of light hitting such a mirror will return along the incident direction.
    Difficulty: 3

MULTIPLE CHOICE

67. Rainbows are caused by sunlight being:
    a) dispersed
    b) refracted
    c) reflected
    d) all of the above

    Answer: d Difficulty: 2

FILL-IN-THE-BLANK

68. Red light with a wavelength of 650.nm travels from air into a liquid with an index of 1.33. What is the frequency and wavelength in the liquid?

    Answer: 4.61×10^{14} Hz and 489.nm Difficulty: 2

MULTIPLE CHOICE

69. Snell's law can be written as: $\sin(\theta_1) / \sin(\theta_2) =$
    a) v_2 / v_1
    b) λ_1 / λ_2
    c) f_1 / f_2
    d) $(\lambda_1 f_2)/(\lambda_2 f_1)$

    Answer: b Difficulty: 2

70. White light is
    a) dichromatic
    b) achromatic
    c) monochromatic
    d) polychromatic

Answer: c Difficulty: 1

FILL-IN-THE-BLANK

71. Arrange in order of increasing wavelength: blue, green, indigo, orange, red, violet, yellow.

Answer: violet, indigo, blue, green, yellow, orange, red. Difficulty: 2

MULTIPLE CHOICE

72. When looking into a pond or lake of clear water, the apparent depth is what fraction of the actual depth?
    a) 1/3
    b) 1/4
    c) 1/2
    d) 3/4
    e) 9/10

Answer: d Difficulty: 2

FILL-IN-THE-BLANK

73. How long does it take a light beam to cross a 6.m wide room? If the air were replaced with a transparent medium of index n, how much longer would it take ?

Answer: 2×10^{-8} s ; n times longer. Difficulty: 2

MULTIPLE CHOICE

74. Which of the folowing is equivalent to Snell's Law? (v is light velocity)
    a) $v_1 / v_2 = \sin(\Theta_1) \sin(\Theta_2)$
    b) $v_1 \tan(\Theta_1) = v_2 \tan(\Theta_2)$
    c) $v_1 / v_2 = \sin(\Theta_1) / \sin(\Theta_2)$
    d) $v_1 / v_2 = \sin(\Theta_2) / \sin(\Theta_1)$
    e) $n_1 \sin(\Theta_2) = n_2 \sin(\Theta_1)$

Answer: c Difficulty: 2

Chapter 22

75. As light refracts going from one medium to another, which of the following remains constant?
   a) refractive index
   b) wave velocity
   c) wavelength
   d) frequency
   e) none of the above

Answer: d Difficulty: 2

Chapter 23

MULTIPLE CHOICE

1. Two light sources are said to be coherent if they
   a) are of the same frequency.
   b) are of the same frequency, and maintain a constant phase difference.
   c) are of the same amplitude, and maintain a constant phase difference.
   d) are of the same frequency and amplitude.

   Answer: b Difficulty: 2

2. Two beams of coherent light travel different paths arriving at point P. If the maximum constructive interference is to occur at point P, the two beams must
   a) arrive 180° out of phase.
   b) arrive 90° out of phase.
   c) travel paths that differ by a whole number of wavelengths.
   d) travel paths that differ by an odd number of half-wavelengths.

   Answer: c Difficulty: 2

3. One beam of coherent light travels path P_1 in arriving at point Q and another coherent beam travels path P_2 in arriving at the same point. If these two beams are to interfere destructively, the path difference $P_1 - P_2$ must be equal to
   a) an odd number of half-wavelengths.
   b) zero.
   c) a whole number of wavelengths.
   d) a whole number of half-wavelengths.

   Answer: a Difficulty: 2

Chapter 23

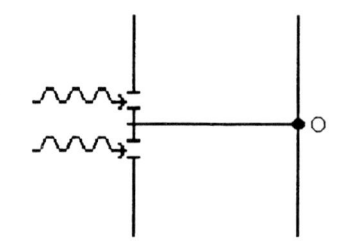

4. Monochromatic light from a distant source is incident on two parallel narrow slits. After passing through the slits the light strikes a screen, as shown in the sketch. What will be the nature of the pattern of light observed on the screen?
   a) The screen will be most brightly illuminated at point O, with the intensity decreasing slowly and uniformly as you move outward from point O.
   b) A series of alternating light and dark bands.
   c) A rainbow of colored lines will be seen spreading out on either side of point O.
   d) Two bright bands of light, one in line with each slit.
   e) The screen will be uniformly illuminated except for two dark bands, one in line with each slit.

   Answer: b Difficulty: 2

5. In a double-slit experiment, it is observed that the distance between adjacent maxima on a remote screen is 1 cm. What happens to the distance between adjacent maxima when the slit separation is cut in half?
   a) It increases to 2 cm.
   b) It increases to 4 cm.
   c) It decreases to 0.5 cm.
   d) It decreases to 0.25 cm.

   Answer: a Difficulty: 2

6. At the first maxima on either side of the central bright spot in a double-slit experiment, light from each opening arrives
   a) in phase.
   b) 90° out of phase.
   c) 180° out of phase.
   d) none of the above.

   Answer: a Difficulty: 2

7. At the second maxima on either side of the central bright spot in a double-slit experiment, light from
   a) each opening travels the same distance.
   b) one opening travels twice as far as light from the other opening.
   c) one opening travels one wavelength of light farther than light from the other opening.
   d) one opening travels two wavelengths of light farther than light from the other opening.

   Answer: d Difficulty: 2

8. The separation between adjacent maxima in a double-slit interference pattern using monochromatic light is
   a) greatest for red light.
   b) greatest for green light.
   c) greatest for blue light.
   d) the same for all colors of light.

   Answer: a Difficulty: 2

ESSAY

9. In a diffraction experiment, light of 600 nm wavelength produces a first-order maximum 0.35 mm from the central maximum on a distant screen. A second monochromatic source produces a third-order maximum 0.87 mm from the central maximum when it passes through the same diffraction grating. What is the wavelength of the light from the second source?

   Answer: 497 nm Difficulty: 2

10. In a double-slit experiment, the slit separation is 2 mm, and two wavelengths, 750 nm and 900 nm, illuminate the slits. A screen is placed 2 m from the slits. At what distance from the central maximum on the screen will a bright fringe from one pattern first coincide with a bright fringe from the other?

    Answer: 4.5 mm Difficulty: 2

11. A single slit, which is 0.05 mm wide, is illuminated by light of 550 nm wavelength. What is the angular separation between the first two minima on either side of the central maximum?

    Answer: 0.630° Difficulty: 2

Chapter 23

MULTIPLE CHOICE

12. If two light waves are coherent, which of the following is NOT necessary?
    a) They must have the same frequency.
    b) They must have the same wavelength.
    c) They must have the same amplitude.
    d) They must have the same velocity.
    e) They must have a constant phase difference at every point in space.

    Answer: c Difficulty: 2

13. Why would it be impossible to obtain interference fringes in a double-slit experiment if the separation of the slits is less than the wavelength of the light used?
    a) The very narrow slits required would generate many different wavelengths, thereby washing out the interference pattern.
    b) The two slits would not emit coherent light.
    c) The fringes would be too close together.
    d) In no direction could a path difference as large as one wavelength be obtained, and this is needed if a bright fringe, in addition to the central fringe, is to be observed.

    Answer: d Difficulty: 2

14. What do we mean when we say that two light rays striking a screen are in phase with each other?
    a) When the electric field due to one is a maximum, the electric field due to the other is also a maximum, and this relation is maintained as time passes.
    b) They are traveling at the same speed.
    c) They have the same wavelength.
    d) They alternately reinforce and cancel each other.

    Answer: a Difficulty: 2

15. The colors on an oil slick are caused by reflection and
    a) diffraction.
    b) interference.
    c) refraction.
    d) polarization.

    Answer: b Difficulty: 2

16. When a beam of light, which is traveling in glass, strikes an air boundary, there is
    a) a 90° phase change in the reflected beam.
    b) no phase change in the reflected beam.
    c) a 180° phase change in the reflected beam.
    d) a 45° phase change in the reflected beam.

    Answer: b Difficulty: 2

17. When a beam of light, which is traveling in air, is reflected by a glass surface, there is
    a) a 90° phase change in the reflected beam.
    b) no phase change in the reflected beam.
    c) a 180° phase change in the reflected beam.
    d) a 45° phase change in the reflected beam.

    Answer: c Difficulty: 2

18. A soap film is being viewed in white light. As the film becomes very much thinner than the wavelength of blue light, the film
    a) appears black because it reflects no visible light.
    b) appears white because it reflects all wavelengths of visible light.
    c) appears blue since all other colors are transmitted.
    d) appears red since all other colors are transmitted.

    Answer: a Difficulty: 2

19. In terms of the wavelength of light in soapy water, what is the minimum thickness of soap film that will reflect a given wavelength of light?
    a) One-fourth wavelength
    b) One-half wavelength
    c) One wavelength
    d) There is no minimum thickness.

    Answer: a Difficulty: 2

20. In terms of the wavelength of light in magnesium fluoride, what is the minimum thickness of magnesium fluoride coating that must be applied to a glass lens to make it non-reflecting for that wavelength? (The index of refraction of magnesium fluoride is intermediate to that of glass and air.)
    a) One-fourth wavelength
    b) One-half wavelength
    c) One wavelength
    d) There is no minimum thickness.

    Answer: a Difficulty: 2

21. A convex lens is placed on a flat glass plate and illuminated from above with monochromatic red light. When viewed from above, concentric bands of red and dark are observed. What does one observe at the exact center of the lens where the lens and the glass plate are in direct contact?
    a) A bright red spot
    b) A dark spot
    c) A rainbow of color
    d) A bright spot that is some color other than red

    Answer: b Difficulty: 2

Chapter 23

22. We have seen that two monochromatic light waves can interfere constructively or destructively, depending on their phase difference. One consequence of this phenomena is
   a) the colors you see when white light is reflected from a soap bubble.
   b) the appearance of a mirage in the desert.
   c) a rainbow.
   d) the way in which polaroid sunglasses work.
   e) the formation of an image by a converging lens, such as the lens in your eye.

   Answer: a Difficulty: 2

ESSAY

23. A soap bubble has an index of refraction of 1.33. What minimum thickness of this bubble will ensure maximum reflectance of normally incident 530 nm wavelength light?

    Answer: 99.6 nm Difficulty: 2

24. A lens is coated with material of index of refraction 1.38. What thickness of coating should be used to give maximum transmission at a wavelength of 530 nm?

    Answer: 96 nm Difficulty: 2

25. Light of wavelength 500 nm illuminates a soap film (n = 1.33). What is the minimum thickness of film that will give an interference when the light is incident normally on it?

    Answer: 94 nm Difficulty: 2

26. A glass plate 2.5 cm long is separated from another glass plate by a strand of someone's hair (diameter 0.001 cm). How far apart are the adjacent interference bands when viewed with light of wavelength 600 nm?

    Answer: 0.75 mm Difficulty: 2

MULTIPLE CHOICE

27. After a rain, one sometimes sees brightly colored oil slicks on the road. These are due to
    a) interference effects.
    b) polarization effects.
    c) diffraction effects.
    d) selective absorption of different wavelengths by oil.
    e) none of the above.

    Answer: a Difficulty: 2

28. Consider two diffraction gratings; one has 4000 lines per cm and the other one has 6000 lines per cm.
   a) The 4000-line grating produces the greater dispersion.
   b) The 6000-line grating produces the greater dispersion.
   c) Both gratings produce the same dispersion, but the orders are sharper for the 4000-line grating.
   d) Both gratings produce the same dispersion, but the orders are sharper for the 6000-line grating.

   Answer: b   Difficulty: 2

29. Consider two diffraction gratings with the same slit separation, the only difference being that one grating has 3 slits and the other 4 slits. If both gratings are illuminated with a beam of the same monochromatic light,
   a) the grating with 3 slits produces the greater separation between orders.
   b) the grating with 4 slits produces the greater separation between orders.
   c) both gratings produce the same separation between orders.
   d) both gratings produce the same separation between orders, but the orders are better defined with the 4-slit grating.

   Answer: d   Difficulty: 2

ESSAY

30. A diffraction grating has 6000 lines per centimeter ruled on it. What is the angular separation between the second and the third orders on the same side of the central order when the grating is illuminated with a beam of light of wavelength 550 nm?

   Answer: 40.5°   Difficulty: 2

31. When red light illuminates a grating with 7000 lines per centimeter, its second maximum is at 62.4°. What is the wavelength of the light?

   Answer: 633 nm   Difficulty: 2

MULTIPLE CHOICE

32. An ideal polarizer is placed in a beam of unpolarized light and the intensity of the transmitted light is 1. A second ideal polarizer is placed in the beam with its referred direction rotated 40° to that of the first polarizer. What is the intensity of the beam after it has passed through both polarizers?
   a) 0.7661
   b) 0.6431
   c) 0.5871
   d) 0.4131

   Answer: c   Difficulty: 2

Chapter 23

33. For a beam of light, the direction of polarization is defined as
    a) the beam's direction of travel.
    b) the direction of the electric field's vibration.
    c) the direction of the magnetic field's vibration.
    d) the direction that is mutually perpendicular to the electric and magnetic field vectors.

    Answer: b Difficulty: 2

34. A polarizer (with its preferred direction rotated 30° to the vertical) is placed in a beam of unpolarized light of intensity 1. After passing through the polarizer, the beam's intensity is
    a) 0.250I
    b) 0.500I
    c) 0.866I
    d) 0.750I

    Answer: b Difficulty: 2

35. What is the Brewster's angle for light traveling in vacuum and reflecting off a piece of glass of index of refraction 1.52?
    a) 48.9°
    b) 41.1°
    c) 33.3°
    d) 56.7°

    Answer: d Difficulty: 2

36. Two emerging beams of light produced by a birefringent crystal
    a) are at different frequencies.
    b) will produce an interference pattern when recombined.
    c) are polarized in the same direction.
    d) are polarized in mutually perpendicular directions.

    Answer: d Difficulty: 2

37. Suppose that you take the lenses out of a pair of polarized sunglasses and place one on top of the other. Rotate one lens 90° with respect to the normal position of the other lens. Early in the morning, look directly overhead at the sunlight coming down. What would you see?
    a) The lenses would look completely dark, since they would transmit no light.
    b) You would see some light, and the brightness would be a function of how the lenses were oriented with respect to the incoming light.
    c) You would see light with intensity reduced to about 50% of what it would be with one lens.
    d) You would see some light, and the intensity would be the same as if you had looked through only one lens.
    e) You would see some light, and the intensity would be greater than if you looked through only one lens.

    Answer: a Difficulty: 2

ESSAY

38. A beam of light passes through a polarizer and then an analyzer. In this process, the intensity of the light transmitted is reduced to 10% of the intensity incident on the analyzer. What is the angle between the axes of the polarizer and the analyzer?

Answer: 71.6° Difficulty: 2

MULTIPLE CHOICE

39. LCDs (Liquid Crystal Displays) such as those on most calculators (probably yours) are based on the physical phenomena called
    a) diffraction.
    b) Rayleigh scattering.
    c) birefringence.
    d) dichroism.
    e) polarization.

Answer: e Difficulty: 2

40. Sunlight reflected from the surface of a lake
    a) is unpolarized.
    b) tends to be polarized with its electric field vector parallel to the surface of the lake.
    c) tends to be polarized with its electric field vector perpendicular to the surface of the lake.
    d) has undergone refraction by the surface of the lake.
    e) none of the above is true.

Answer: b Difficulty: 2

41. The polarization of sunlight is greatest at
    a) sunrise.
    b) sunset.
    c) both sunrise and sunset.
    d) midday.

Answer: c Difficulty: 2

42. If sunlight of color B is scattered through an angle 16 times greater than sunlight of color A, then the wavelength of color B is
    a) twice that of color A.
    b) 1/2 that of color A.
    c) 16 times that of color A.
    d) 1/16 that of color A.
    e) none of the above.

Answer: b Difficulty: 2

Chapter 23

43. On a clear day, the sky appears to be more blue toward the zenith (overhead) than it does toward the horizon. This occurs because
    a) the atmosphere is denser higher up than it is at the earth's surface.
    b) the temperature of the upper atmosphere is higher than it is at the earth's surface.
    c) the sunlight travels over a longer path at the horizon, resulting in more absorption.
    d) none of the above is true.

    Answer: c   Difficulty: 2

FILL-IN-THE-BLANK

44. A diver looks in a mirror **under water**. If the diver's face is 30.cm from the mirror and the water index is 1.333, what is the lateral magnification?

    Answer: ONE; the index has no effect upon REFLECTION.   Difficulty: 2

MULTIPLE CHOICE

45. A light pipe (e.g. glass fiber) works because of:
    a) dispersion
    b) interference
    c) total internal refraction
    d) total internal reflection
    e) the circular cross section area

    Answer: d   Difficulty: 2

46. Which of the following aberrations does not apply to mirrors?
    a) spherical
    b) chromatic
    c) astigmatism
    d) coma

    Answer: b   Difficulty: 2

FILL-IN-THE-BLANK

47. Is it possible for a convex mirror to form a real image?

    Answer: Yes, if the object was virtual (formed by another lens or mirror)

    Difficulty: 2

TRUE/FALSE

48. The radius of curvature for a plane (flat) mirror is zero.

    Answer: F   Difficulty: 1

Testbank

MULTIPLE CHOICE

49. A fish appears to be at the center of a spherical fish bowl. We know light rays leaving the bowl refract at the glass surface so where actually is the fish? (Assume the water and glass have n=1.33)
   a) 1.33R behind the glass
   b) at the center
   c) half way to the center
   d) R/1.33 behind the glass

   Answer: b Difficulty: 2

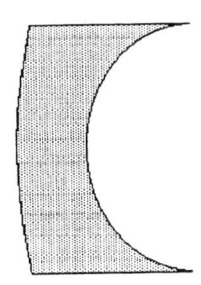

50. A concave meniscus lens with an index of refraction of 1.48 has radii of 6.0 cm and 4.0 cm (see figure). What is the power of the lens in DIOPTERS?
   a) -5.
   b) -0.2
   c) -0.04
   d) -0.25
   e) -4.
   f) -20.
   g) -12.

   Answer: e Difficulty: 2

51. In air a concave mirror has a focal length of 30. cm. Under water that same mirror would have _____ focal length.
   a) a larger
   b) the same
   c) a smaller

   Answer: b Difficulty: 2

52. If the refractive index of the glass in a lens varies with wavelength, then the following results:
   a) spherical aberration
   b) light scattering
   c) total reflection
   d) astimatism
   e) chromatic aberration

   Answer: e Difficulty: 2

361

Chapter 23

ESSAY

53. Take the lens maker equation and show that a lens which has been turned around (surface 1 becomes 2 and 2 becomes 1) as the same focal length.

    Answer:
    $1/f = (n-1)(1/R_1 - 1/R_2)$ A sign change occurs when 1 and 2 are exchanged, and the radii themselves change sign because the center of curvature moves to other side of surface. These sign changes cancel each other out and yield the same f.
    Difficulty: 2

MULTIPLE CHOICE

54. What kind of lens is it that has a power of +10 Diopters and a first surface center of curvature in front of the lens?
    a) double convex
    b) double concave
    c) converging meniscus
    d) diverging meniscus
    e) plano concave
    f) plano convex

    Answer: c Difficulty: 2

55. Two parallel rays (also parallel to the optic axis) reflect from a concave mirror. They intersect at:
    a) the center of curvature
    b) the focal point
    c) a point behind the mirror
    d) the image point

    Answer: d Difficulty: 2

FILL-IN-THE-BLANK

56. The curved surface, of a 50.cm focal length "plano-convex" lens, has a radius of curvature of 30.cm. What is the refractive index of the glass?

    Answer: n = 1.6 Difficulty: 2

57. A 4.0 cm high object is 20.cm in front of a 1.333 Diopter plano-convex lens.
    a) Find the image position. Real or Virtual?
    b) Image size? Erect Inverted?

    Answer: a) 60.cm in front of lens; virtual image   b) 12.cm high; erect

    Difficulty: 2

MULTIPLE CHOICE

58. What is the FOCAL LENGTH of a lens whose two radii of curvatures are equal ($R_1 = R_2$)?
   a) $2R_2/(n-1)$
   b) zero
   c) infinite
   d) $R_1/(2n-2)$
   e) none of the above

   Answer: c   Difficulty: 2

FILL-IN-THE-BLANK

59. A slide projector is designed with its lens 6.0 m from the screen. The projected image is 1.5 m square of a slide object 2.5 cm square.
   a) What is the object distance from the lens?
   b) What should be the lens focal length?
   c) What is the lens POWER?

   Answer: a) 10. cm   b) f=9.8 cm   c) 10.2 Diopters   Difficulty: 2

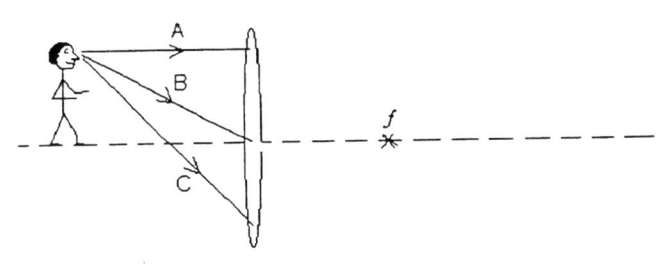

60. A lens projects an image of a man as seen in the figure. Rays marked A, B, and C travel to the lens from the man's ear. Draw the three refracted rays as they proceed to the right of the lens; noting that A is parallel to the axis, B goes through the center of the lens, C proceeds to the bottom of the lens, and the point marked f is the focal point of the lens.

   Answer: (see figure)

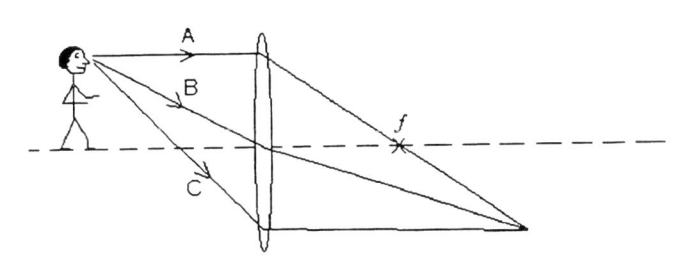

   Difficulty: 2

Chapter 23

MULTIPLE CHOICE

61. Consider a converging lens. If the object is inside the focal point, then the image will always be
    a) real
    b) virtual
    c) at infinity
    d) at the focal point

    Answer: b Difficulty: 2

62. The power of the human eye is approximately how many Diopters ? (hint: estimate the size of the eyeball)
    a) -25
    b) -4
    c) -1
    d) 1
    e) 4
    f) 25

    Answer: f Difficulty: 2

Chapter 24

MATCHING

Match the eye-part's function to its name.

a) a circular diaphragm
b) the central hole
c) a light-sensitive surface, upon which the image falls
d) a curved transparent tissue where light first enters the eye
e) composed of microscopic glassy fibers that slide over each other
f) relays signals to the brain
g) photosensitive receptor cells of which the retina is composed
h) a jelly-like substance that fills most of the eye

1. iris

    Answer: a) a circular diaphragm Difficulty: 2

2. pupil

    Answer: b) the central hole Difficulty: 2

3. retina

    Answer: c) a light-sensitive surface, upon which the image falls

    Difficulty: 2

4. cornea

    Answer: d) a curved transparent tissue where light first enters the eye

    Difficulty: 2

5. lens

    Answer: e) composed of microscopic glassy fibers that slide over each other

    Difficulty: 2

6. optic nerve

    Answer: f) relays signals to the brain Difficulty: 2

7. rods and cones

    Answer: g) photosensitive receptor cells of which the retina is composed

    Difficulty: 2

Chapter 24

8. vitreous humor

   Answer: h) a jelly-like substance that fills most of the eye

   Difficulty: 2

TRUE/FALSE

9. The aperture in a camera plays the same role as the retina in the eye.

   Answer: F Difficulty: 2

10. 30/40 vision means that one eye can see clearly at 30 ft, and the other eye can see clearly at 40 ft.

    Answer: F Difficulty: 2

MULTIPLE CHOICE

11. What power lens is needed to correct for nearsightedness where the uncorrected far point is 75 cm?
    a) -0.75 D
    b) +1.33 D
    c) -1.33 D
    d) None of the above

    Answer: c Difficulty: 2

12. What power lens is needed to correct for farsightedness where the uncorrected near point is 75 cm?
    a) -2.67 D
    b) +2.67 D
    c) +5.33 D
    d) None of the above

    Answer: b Difficulty: 2

ESSAY

13. A nearsighted man wears -3.0 D lenses. With these lenses, his corrected near point is 25 cm. What is his uncorrected near point?

    Answer: 14.3 cm Difficulty: 2

14. A nearsighted person has a near point of 12 cm and a far point of 17 cm. If the lens is 2 cm from the eye, (a) what lens power will enable this person to see distant objects clearly? (b) what then will be the new near point?

    Answer: (a) -6.7 D   (b) 30 cm in front of the lens Difficulty: 2

15. A person uses corrective glasses of power -8.5 D. (a) Is this person nearsighted or farsighted? (b) If the glasses are worn 2 cm from the eye, what is the person's far point without glasses?

Answer: (a) nearsighted   (b) 14 cm   Difficulty: 2

MULTIPLE CHOICE

16. A fish's eye is well-suited to seeing under water, but what can you say about his vision if he is taken out of water?
    a) His acuity (sharpness of vision) would be greater.
    b) He would suffer from astigmatism.
    c) He would be nearsighted.
    d) He would be farsighted.
    e) His vision would be limited to a cone of half-angle of about 49°.

    Answer: c Difficulty: 2

17. The principal refraction of light by the eye occurs at the
    a) cornea.
    b) lens.
    c) retina.
    d) sclera.
    e) iris.

    Answer: a Difficulty: 2

18. Rays that pass through a lens very close to the lens axis are more sharply focused than those that are very far from the axis. The inability of a lens to sharply focus off-axis rays is called spherical aberration. This effect helps us to understand why
    a) we become more farsighted as we grow older.
    b) we become more nearsighted as we grow older.
    c) we can see only in black and white in dim light.
    d) it is easier to read in bright light than in dim light.
    e) a moving object is more readily detected than a stationary one.

    Answer: d Difficulty: 2

19. If a person's eyeball is too short from front to back, the person is likely to suffer from
    a) astigmatism.
    b) spherical aberration.
    c) farsightedness.
    d) nearsightedness.

    Answer: c Difficulty: 2

Chapter 24

20. If a person's eyeball is too long from front to back, the person is likely to suffer from
    a) spherical aberration.
    b) nearsightedness.
    c) farsightedness.
    d) astigmatism.

    Answer: b   Difficulty: 2

21. Nearsightedness can usually be corrected with
    a) converging lenses.
    b) diverging lenses.
    c) achromatic lenses.
    d) cylindrical lenses.

    Answer: b   Difficulty: 2

22. The near point of a farsighted person is 100 cm. She places reading glasses close to her eye, and with them she can comfortably read a newspaper at a distance of 25 cm. What lens power is required?
    a) +2.5 D
    b) +3.0 D
    c) +3.2 D
    d) -2.0 D

    Answer: b   Difficulty: 2

23. Farsightedness can usually be corrected with
    a) cylindrical lenses.
    b) achromatic lenses.
    c) diverging lenses.
    d) converging lenses.

    Answer: d   Difficulty: 2

24. If the human eyeball is too short from front to back, this gives rise to a vision defect that can be corrected by using
    a) convex-convex eyeglasses.
    b) concave-convex eyeglasses.
    c) cylindrical eyeglasses.
    d) contact lenses, but no ordinary lenses.
    e) shaded glasses (i.e., something that will cause the iris to dilate more).

    Answer: a   Difficulty: 2

25. In which of the following ways is a camera different from the human eye?
    a) The camera always forms an inverted image, the eye does not.
    b) The camera always forms a real image, the eye does not.
    c) The camera utilizes a fixed focal length lens, the eye does not.
    d) For the camera, the image magnification is greater than one, but for the eye the magnification is less than one.
    e) A camera cannot focus on objects at infinity but the eye can.

    Answer: c Difficulty: 2

ESSAY

26. What is the maximum angular magnification of a magnifying glass of focal length 10 cm? (Assume the near point is at 25 cm.)

    Answer: 2.5 Difficulty: 2

27. A magnifying glass of focal length 15 cm is used to examine an old manuscript. (a) What is the maximum magnification given by the lens? (b) What is the magnification for relaxed eye-viewing (image at infinity)?

    Answer: (a) 2.67   (b) 1.67 Difficulty: 2

MULTIPLE CHOICE

28. The first important discoveries made with a telescope (in 1609) were done by
    a) Hale.
    b) Newton.
    c) Cassegrain.
    d) Galileo.

    Answer: d Difficulty: 2

29. All of the following are reflecting telescopes, except
    a) Newtonian.
    b) Yerkes.
    c) Cassegrain.
    d) Hale.

    Answer: b Difficulty: 2

30. The objective of a telescope has a focal length of 100 cm and its eyepiece has a focal length of 5 cm. What is the magnification of this telescope when viewing an object at infinity?
    a) 20
    b) 0.05
    c) 500
    d) None of the above

    Answer: a Difficulty: 2

Chapter 24

ESSAY

31. A refracting telescope has an eyepiece of focal length -8 cm and an objective of focal length 36 cm. What is the magnification?

    Answer: 4.5 x Difficulty: 2

32. A student wishes to build a telescope. She has available an eyepiece of focal length of -5 cm. What focal length objective is needed to obtain a magnification of 10x?

    Answer: 50 cm Difficulty: 2

33. An objective lens of a telescope has a 1.9-m focal length. When viewed through this telescope, the moon appears 5.25 times larger than normal. How far apart are the objective lens and the eyepiece when this instrument is focused on the moon?

    Answer: 2.26 m Difficulty: 2

MULTIPLE CHOICE

34. Consider the image formed by a refracting telescope. Suppose an opaque screen is placed in front of the lower half of the objective lens. What effect will this have?
    a) The top half of the image will be blacked out.
    b) The lower half of the image will be blacked out.
    c) The entire image will be blacked out, since the entire lens is needed to form an image.
    d) The image will appear as it would if the objective were not blocked, but it will be dimmer.
    e) There will be no noticeable difference in the appearance of the image with the objective partially blocked or not.

    Answer: d Difficulty: 2

35. If the diameter of a radar dish is doubled, what happens to its resolving power assuming that all other factors remain unchanged? Its resolving power
    a) quadruples.
    b) doubles.
    c) halves.
    d) is reduced to one-quarter its original value.

    Answer: b Difficulty: 2

ESSAY

36. A single slit, which is 0.05 mm wide, is illuminated by light of 550-nm wavelength. What is the angular separation between the first two minima on either side of the central maximum?

    Answer: 0.630° Difficulty: 2

37. What is the smallest angular separation (in minutes of arc) for which two point sources can be resolved using 3-cm microwaves and a 3-m diameter radar dish?

    Answer: 41.9' Difficulty: 2

MULTIPLE CHOICE

38. In which of the following is diffraction NOT exhibited?
    a) Viewing a light source through a small pinhole
    b) Examining a crystal by X-rays
    c) Using a microscope under maximum magnification
    d) Resolving two nearby stars with a telescope
    e) Determining the direction of polarization with a birefringent crystal

    Answer: e Difficulty: 2

39. In single-slit diffraction
    a) the intensities of successive maxima are roughly the same, falling off only gradually as one goes away from the central maximum.
    b) the central maximum is about as wide as the other maxima.
    c) the central maximum is about twice as wide as the other maxima.
    d) the central maximum is much narrower than the other maxima.
    e) the slit width must be less than one wavelength for a diffraction pattern to be apparent.

    Answer: c Difficulty: 2

ESSAY

40. The world's largest refracting telescope is operated at the Yerkes Observatory in Wisconsin. It has an objective of diameter of 102 cm. Suppose such an instrument could be mounted on a spy satellite at an elevation of 300 km above the surface of the earth. What is the minimum separation of two objects on the ground if their images are to be clearly resolved by this lens? (Assume an average wavelength of 550 nm for white light.)

    Answer: 0.20 m Difficulty: 2

Chapter 24

MULTIPLE CHOICE

41. With what color light would you expect to be able to see the greatest detail when using a microscope?
    a) red, because of its long wavelength
    b) red, because it is refracted less than other colors by glass
    c) blue, because of its shorter wavelength
    d) blue, because it is brighter
    e) The color makes no difference in the resolving power, since this is determined only by the diameter of the lenses.

    Answer: c Difficulty: 2

42. An important reason for using a very large diameter objective in an astronomical telescope is
    a) to increase the magnification.
    b) to increase the resolution.
    c) to form a virtual image, which is easier to look at.
    d) to increase the width of the field of view.
    e) to increase the depth of the field of view.

    Answer: b Difficulty: 2

43. Radio waves are diffracted by large objects such as buildings, whereas light is not. Why is this?
    a) Radio waves are unpolarized, whereas light is plane polarized.
    b) The wavelength of light is much smaller than the wavelength of radio waves.
    c) The wavelength of light is much greater than the wavelength of radio waves.
    d) Radio waves are coherent, and light is usually not coherent.

    Answer: b Difficulty: 2

44. A person gazes at a very distant light source. If she now holds up two fingers, with a very small gap between them, and looks at the light source, she will see
    a) the same thing as without the fingers, but dimmer.
    b) a series of bright spots.
    c) a sequence of closely spaced bright lines.
    d) a hazy band of light varying from red at one side to blue or violet at the other.

    Answer: c Difficulty: 2

ESSAY

45. Railroad tracks are spaced about 1.7 m apart. On a clear day, how high could a plane fly before the pilot is no longer able to resolve the two separate rails? (Assume that the diameter of the pupil is about 3 mm and take as the wavelength that to which the eye is most sensitive, namely 530 nm.)

    Answer: About 8 km Difficulty: 2

46. If the pupil diameter of the dark-adapted eye is 5 mm, what is the maximum distance at which two point sources of light that are 3 mm apart can still be resolved? (You might try to confirm this by using a cardboard mask with two small pinholes placed in front of a light source that is completely enclosed.)

    Answer: 25 m Difficulty: 2

MULTIPLE CHOICE

47. Assuming the film used has uniform sensitivity throughout the visible spectrum, in which of the following cases would you be able to best distinguish between two closely spaced stars? (The lens referred to is the objective lens of the telescope used.)
    a) Use a large lens and blue light.
    b) Use a large lens and red light.
    c) Use a small lens and blue light.
    d) Use a small lens and red light.

    Answer: a Difficulty: 2

48. The resolving power of a microscope refers to the ability to
    a) distinguish objects of different colors.
    b) form clear images of two points that are very close together.
    c) form a very large image.
    d) form a very bright image.

    Answer: b Difficulty: 2

49. Color blindness is caused by
    a) damage to the retina.
    b) cataracts.
    c) a missing cone-type.
    d) lack of melanin.
    e) too small an iris.

    Answer: c Difficulty: 2

Chapter 24

50. Black light
    a) has a frequency higher than violet light.
    b) has a frequency between red light and violet light.
    c) has a frequency lower than red light.
    d) is not a color of light.

    Answer: d Difficulty: 2

51. Color TV is produced by
    a) red, green, and blue beams of electrons merging at the screen.
    b) cyan, magenta, and yellow beams of electrons merging at the screen.
    c) A single beam of electrons striking a triad of phosphors of red, green, and blue.
    d) A single beam of electrons striking a triad of phosphors of cyan, magenta, and yellow.

    Answer: c Difficulty: 2

52. If a red filter were placed in front of blue dungarees, you would see
    a) red.
    b) blue.
    c) green.
    d) white.
    e) black.

    Answer: e Difficulty: 2

53. If you wear green sunglasses, and look at equations written in white chalk on a chalkboard, you would see the chalk as
    a) green.
    b) white.
    c) black.
    d) red.
    e) yellow.

    Answer: a Difficulty: 2

ESSAY

54. Discuss a similarity and a difference between the spectrum formed by a diffraction grating (a large number of slits) and that formed by just a double slit.

    Answer:
    The maxima of both satisfy $d \sin(\theta) = n\lambda$, but the fringes formed by the grating are much sharper.
    Difficulty: 2

TRUE/FALSE

55. Another name for the POLARIZING ANGLE is the BREWSTER ANGLE.

    Answer: T Difficulty: 1

56. The fact that the sky appears blue is due to polarization.

    Answer: F Difficulty: 1

57. The path difference for destructive interference is $m\lambda/2$ where $m=0,2,4,6,\ldots$

    Answer: F Difficulty: 1

MULTIPLE CHOICE

58. Light which reflects off a pond of water undergoes a phase shift of:
    a) $0.°$
    b) $90.°$
    c) $180.°$
    d) $270.°$

    Answer: c Difficulty: 2

59. Constructive interference of two coherent waves will occur if the path difference is:
    a) $3/2\ \lambda$
    b) $6/2\ \lambda$
    c) $3/4\ \lambda$
    d) none of the above

    Answer: b Difficulty: 2

ESSAY

60. Why does light seem to go straight through an open doorway whereas sound seems to spread out when it goes through the same doorway?

    Answer:
    It has to do with the relative sizes of their wavelengths. Wavelength of light is << opening size so very small diffraction; sound wavelengths >> opening so diffraction is significant.

    Difficulty: 2

Chapter 24

MULTIPLE CHOICE

61. Liquid Crystal Displays (LCDs) depend upon which of the following for their operation?
    a) diffraction
    b) polarization
    c) coherence
    d) dichroism
    e) Bragg scattering
    f) thin film interference

    Answer: b Difficulty: 1

62. How many orders are in the spectra formed by a grating with 3000.lines/cm illuminated by red light of wavelength 600.nm?
    a) 0
    b) 1
    c) 2
    d) 3
    e) 4
    f) 5

    Answer: f Difficulty: 2

63. The condition $2d \sin(\theta) = n\lambda$ for X-ray diffraction maxima is attributed to:
    a) Young
    b) Land
    c) Bragg
    d) Brewster
    e) Rayleigh

    Answer: c Difficulty: 1

64. Rayleigh scattering would have 400.nm light be scattered how many times as much as 800.nm light?
    a) 1
    b) 2
    c) 4
    d) 8
    e) 16
    f) 32

    Answer: e Difficulty: 2

65. A material which has the ability to rotate the direction of polarization of linearly polarized light is said to be:
   a) birefringent
   b) dichroic
   c) optically active
   d) circularly polarized
   e) diffraction limited

   Answer: c Difficulty: 2

66. A ray of light traveling in water (n=1.33) hits a glass surface (n=1.50). At what angle, with the surface, must the incident ray be in order that the polarization of the reflected ray is the greatest?
   a) 56.3°
   b) 33.7°
   c) 48.4°
   d) 41.6°
   e) 53.1°
   f) 36.9°

   Answer: d Difficulty: 2

67. Rayleigh scattering would have 400.nm light be scattered how many times as much as 800.nm light?
   a) 1
   b) 2
   c) 4
   d) 8
   e) 16
   f) 32

   Answer: e Difficulty: 2

68. The sky appears "blue" because of:
   a) birefringence
   b) Rayleigh Scattering
   c) Brewster's polarization
   d) dichroism
   e) Bragg's Law

   Answer: b Difficulty: 2

Chapter 24

ESSAY

69. Discuss what you see viewing a distant street light through a thin curtain (at night).

    Answer:
    One expects to see a "cross" pattern of light which turns out to be the diffraction pattern of the many crossed threads forming multiple two dimensional openings. Note that where the curtain is folded at an angle the pattern enlarges because the projected opening in the line of sight is less and the diffracted angle larger. Close attention will show a colored pattern since large wavelengths are diffracted greater.
    Difficulty: 2

MULTIPLE CHOICE

70. At what angle, with the surface, must light stike glass (refractive index 1.6) in order to completely polarize the reflected ray?
    a) 4.0°
    b) 34.°
    c) 56.°
    d) 86.°

    Answer: a   Difficulty: 2

71. Light in a frozen block of ice reflects off the ice-air interface and undergoes a phase shift of:
    a) 0°
    b) 90°
    c) 180°
    d) 270°

    Answer: a   Difficulty: 2

ESSAY

72. How do interference and diffraction differ? How are they similar?

    Answer:
    Interference is the addition of a finite number (usually small) of coherent waves whereas diffraction involves an infinite (uncountably large) number of waves.
    Difficulty: 2

Chapter 25

ESSAY

1. A boat can travel 4 m/s in still water. With what speed, relative to the shore, does it move in a river that is flowing at 1 m/s if the boat is heading (a) upstream. (b) downstream. (c) straight across the river.

   Answer: (a) 3 m/s   (b) 5 m/s   (c) 3.87 m/s   Difficulty: 2

MULTIPLE CHOICE

2. The Michelson-Morley experiment was designed to measure
   a) the relativistic mass of the electron.
   b) the relativistic energy of the electron.
   c) the velocity of the earth relative to the ether.
   d) the acceleration of gravity on the earth's surface.

   Answer: c   Difficulty: 2

3. Michelson and Morley concluded from the results of their experiment that
   a) the experiment was a failure since there was no detectable shift in the interference pattern.
   b) the experiment was successful in not detecting a shift in the interference pattern.
   c) the experiment was a failure since they detected a shift in the interference pattern.
   d) the experiment was successful in detecting a shift in the interference pattern.

   Answer: a   Difficulty: 2

4. You can build an interferometer yourself if you use the following components:
   a) a light source, a detector screen, a partially silvered mirror, a flat mirror, and a glass plate.
   b) a light source, a detector screen, two partially silvered mirrors, and a glass plate.
   c) a light source, a detector screen, two partially silvered mirrors, a flat mirror, and a glass plate.
   d) a light source, a detector screen, a partially silvered mirror, two flat mirrors, and a glass plate.
   e) a light source, a detector screen, two partially silvered mirrors, two flat mirrors, and a glass plate.

   Answer: d   Difficulty: 2

Chapter 25

TRUE/FALSE

5. One consequence of Einstein's theory of special relativity is that absolute velocity of stars can be measured with respect to the ether wind.

    Answer: F   Difficulty: 2

MULTIPLE CHOICE

6. The gamma factor is defined as $\gamma \equiv 1/\sqrt{1 - (v/c)^2}$, therefore gamma ($\gamma$)
    a) can be zero.
    b) can be a negative number.
    c) can be any number greater than or equal to zero.
    d) can be any number greater than or equal to one.
    e) cannot equal one.

    Answer: d   Difficulty: 2

ESSAY

7. How fast would a rocket ship have to move to contract to half of its proper length (as observed by a stationary object)?

    Answer: 0.87 c   Difficulty: 2

8. A radar operator on earth sees two spaceships moving straight at each other, each with speed 0.6 c. With what speed does the pilot of one ship see the other ship approaching?

    Answer: 0.88 c   Difficulty: 2

9. How fast should a moving clock travel if it is to be observed by a stationary observer as running at one-half its normal rate?

    Answer: 0.866 c   Difficulty: 2

10. A spaceship visits Alpha Centauri and returns to earth. Alpha Centauri is 4.5 light-years from earth (our second closest star). If the spaceship travels at one-half the speed of light for essentially all of its expedition, how long was the ship gone according to an observer on the earth?

    Answer: 18 yrs   Difficulty: 2

11. A spaceship visits Alpha Centauri and returns to earth. Alpha Centauri is 4.5 light-years from earth (our second closest star). If the spaceship travels at one-half the speed of light for essentially all of its expedition, how long was the ship gone according to an observer on the spaceship?

    Answer: 15.6 yrs Difficulty: 2

12. Two spaceships approach each other along a straight line at a constant velocity of 0.988 c as measured by the captain of one of the ships. An observer on earth is able to measure the speed of only one of the ships as 0.900 c. From the point of view of the observer on earth, what is the speed of the other ship?

    Answer: 0.794 c Difficulty: 2

13. Two spaceships are traveling through space at velocities of 0.6 c and 0.9 c, respectively, with respect to earth. If they are headed directly toward each other, what is their approach velocity, as measured by the captain of either ship?

    Answer: 0.974 c Difficulty: 2

14. A person in a rocket ship traveling past the earth at a speed of 0.5 c fires a laser gun in the forward direction. With what speed does an observer on earth see the light pulse travel?

    Answer: 3×10^8 m/s Difficulty: 2

MULTIPLE CHOICE

15. Two spaceships are traveling through space at 0.6 c relative to the earth. If the ships are headed directly toward each other, what is their approach velocity, as measured by a person on either craft?
    a) 1.2 c
    b) c
    c) 0.6 c
    d) None of the above

    Answer: d Difficulty: 2

16. The theory of special relativity
    a) is based on a complex mathematical analysis.
    b) has not been verified by experiment.
    c) does not agree with Newtonian mechanics.
    d) does not agree with electromagnetic theory.

    Answer: c Difficulty: 2

Chapter 25

17. Consider two spaceships, each traveling at 0.5 c in a straight line. Ship A is moving directly away from the sun and ship B is approaching the sun. The science officers on each ship measure the velocity of light coming from the sun. What do they measure for this velocity?
    a) Ship A measures it as less than c, and ship B measures it as greater than c.
    b) Ship B measures it as less than c, and ship A measures it as greater than c.
    c) On both ships it is measured to be less than c.
    d) On both ships it is measured to be exactly c.
    e) On both ships it is measured to be greater than c.

    Answer: d Difficulty: 2

18. Relative to a stationary observer, a moving clock
    a) always runs slower than normal.
    b) always runs faster than normal.
    c) keeps its normal time.
    d) can do any of the above. It depends on the relative velocity between the observer and the clock.

    Answer: a Difficulty: 2

19. As the velocity of your spaceship increases, you would observe
    a) that your precision clock runs slower than normal.
    b) that the length of your spaceship has decreased.
    c) that your mass has increased.
    d) all of the above.
    e) none of the above.

    Answer: e Difficulty: 2

20. You are riding in a spaceship (lucky you!) that has no windows or other means for you to observe or measure what is outside. You wish to determine if the ship is stopped or moving at constant velocity. What should you do?
    a) You can determine if the ship is moving by determining the apparent velocity of light.
    b) You can determine if the ship is moving by checking your precision time piece. If its running slow, the ship is moving.
    c) You can determine if the ship is moving either by determining the apparent velocity of light or by checking your precision time piece. If its running slow, the ship is moving.
    d) You should give up because you have taken on an impossible task.

    Answer: d Difficulty: 2

21. An object moves in a direction parallel to its length with a velocity that approaches the velocity of light. The length of this object, as measured by a stationary observer,
    a) approaches infinity.
    b) approaches zero.
    c) increases slightly.
    d) does not change.

    Answer: b Difficulty: 2

22. An object moves in a direction parallel to its length with a velocity that approaches the velocity of light. The width of this object, as measured by a stationary observer,
    a) approaches infinity.
    b) approaches zero.
    c) increases slightly.
    d) does not change.

    Answer: d Difficulty: 2

23. An object moves in a direction parallel to its length with a velocity that approaches the velocity of light. The mass of this object, as measured by a stationary observer,
    a) approaches infinity.
    b) approaches zero.
    c) increases slightly.
    d) does not change.

    Answer: a Difficulty: 2

24. A spear is thrown at you at a very high speed. As it passes, you measure its length at one-half its normal length. From this measurement, you conclude that the moving spear's mass must be
    a) one-half its rest mass.
    b) twice its rest mass.
    c) four times its rest mass.
    d) none of the above.

    Answer: b Difficulty: 2

25. One of Einstein's postulates in formulating the special theory of relativity was that the laws of physics are the same in reference frames that
    a) accelerate.
    b) move at constant velocity with respect to an inertial frame.
    c) oscillate.
    d) are stationary, but not in moving frames.

    Answer: b Difficulty: 2

Chapter 25

26. If the velocity of your spaceship goes from 0.3 c to 0.6 c, then your mass will increase by
   a) 19%
   b) 38%
   c) 100%
   d) 200%

   Answer: a Difficulty: 2

ESSAY

27. How fast must something be traveling if its mass increases by 10%?

   Answer: 1.25×10^8 m/s Difficulty: 2

MULTIPLE CHOICE

28. If you were to measure your pulse rate while in a spaceship moving away from the sun at a speed close to the speed of light, you would find that it was
   a) much faster than normal.
   b) much slower than normal.
   c) the same as it was here on earth.

   Answer: c Difficulty: 2

ESSAY

29. The atomic bomb that was dropped on Nagasaki in 1945 killed 140,000 people, helping to end World War II on the next day. It released energy equivalent to that of 20,000 tons of TNT explosive. How much mass was converted to energy when this took place? (1000 tons ↔ 4.3×10^{12} J) Incidentally, modern H-bombs have energy yields 1000 times as much!

   Answer: 0.96 g Difficulty: 2

30. A person of initial mass 70 kg climbs a stairway, rising 6 m in elevation. By how much does his mass increase by virtue of his increased potential energy?

   Answer: 4.57×10^{-14} kg Difficulty: 2

MULTIPLE CHOICE

31. Which of the following depends on the observer's frame of reference?
   a) The mass of the proton
   b) The length of a meter stick
   c) The half-life of a muon
   d) All of the above

   Answer: d Difficulty: 2

32. George Gamow, the creator of the Big-Bang theory of the origin of the universe, asks you to imagine what would happen if the speed of light was 15 mi/h. If you were on the sidewalk looking at a bicyclist ride down the street, how does she look to you while riding, compared to when she stops?
   a) taller, the same mass, but stretched out like a limousine.
   b) shorter, the same mass, but stretched out like a limousine.
   c) same height, greater mass, and flatter in the direction she is moving.
   d) same height, smaller mass, and flatter in the direction she is moving.
   e) shorter, greater mass, but stretched out like a limousine.

   Answer: c   Difficulty: 2

TRUE/FALSE

33. The total energy of a particle at rest is zero.

   Answer: F   Difficulty: 2

MULTIPLE CHOICE

34. Consider a particle of mass m and rest mass m_0. Which of the following is the correct expression for the kinetic energy of such a particle?
   a) $m_0 v^2 / 2$
   b) $mv^2 / 2$
   c) $mc^2 - m_0 c^2$
   d) $1/2(mc^2 - m_0 c^2)$

   Answer: c   Difficulty: 2

35. How many joules of energy are required to accelerate one kilogram of mass from rest to a velocity of 0.886 c?
   a) 1.8×10^{17} J
   b) 9×10^{16} J
   c) 3×10^3 J
   d) None of the above

   Answer: a   Difficulty: 2

36. The amount of energy equivalent to one kilogram of mass at rest is
   a) 9×10^{16} J
   b) 3×10^8 J
   c) 4.5×10^{16} J
   d) None of the above.

   Answer: a   Difficulty: 2

Chapter 25

37. What happens to the kinetic energy of a speedy proton when its relativistic mass doubles?
   a) It doubles.
   b) It more than doubles.
   c) It less than doubles.
   d) It must increase, but it is impossible to say by how much.

   Answer: b Difficulty: 2

38. What happens to the total relativistic energy of a speedy proton when its relativistic mass doubles?
   a) It doubles.
   b) It more than doubles.
   c) It less than doubles.
   d) It must increase, but it is impossible to say by how much.

   Answer: a Difficulty: 2

39. An electron is accelerated through 100 kV. By what factor has its mass increased with respect to its rest mass?
   a) 1.20
   b) 1.55
   c) 4.25
   d) 8.00

   Answer: a Difficulty: 2

ESSAY

40. An electron has a relativistic momentum of 1.1×10^{-21} kg-m/s. What fraction of its total energy is its kinetic energy?

    Answer: 0.75 Difficulty: 2

TRUE/FALSE

41. According to Einstein's general theory of relativity, rotating reference frames cannot be distinguished from gravitational acceleration.

    Answer: F Difficulty: 2

MULTIPLE CHOICE

42. The gravitational red shift is caused by
    a) gravitational lensing.
    b) Rayleigh scattering in the atmosphere.
    c) time dilation.
    d) rotating black holes.

    Answer: c Difficulty: 2

43. Black holes
   a) are holes in space, devoid of matter.
   b) are predicted by Einstein's special theory of relativity.
   c) are the collapsed remnant of stars.
   d) cannot be detected in binary star systems.
   e) are a violation of Einstein's general theory of relativity.

   Answer: c Difficulty: 2

44. The Schwarzchild radius of a black hole is that radial distance from the center of a sphere within which not even light can escape. It was first discovered mathematically by Schwarzchild in 1916 after Einstein published his general relativity theory. It can be calculated from a star's mass M as: $R = 2GM/c^2$. If the mass of star G is twice as much as the mass of star H, the average density of star G, compared to star H will be
   a) the same.
   b) twice as much.
   c) half as much.
   d) four times as much.
   e) one-fourth as much.

   Answer: d Difficulty: 2

FILL-IN-THE-BLANK

45. Lisa Anne is nearsighted with a near point of 75. cm.

   a) What POWER lens is needed to correct her nearsightedness (assuming distance from lens to eye is negligible)? _____

   b) What FOCAL LENGTH is this? _____

   Answer: a) -1.33 Diopters,   b) -75. cm   Difficulty: 2

46. A hyperopic eye can focus on objects which are 75. cm and farther. What POWER and what FOCAL LENGTH spectacles should be prescribed so it can see objects 25. cm and farther from the eye? Assume negligible distance between spectacles and eye.

   Answer: POWER = 2.7 Diopters ; f = 38. cm. Difficulty: 2

Chapter 25

MULTIPLE CHOICE

47. Doubling the f-number of a camera lens (e.g. from f/16 to f/32) will change the intensity of light received by the film by what factor?
    a) 4
    b) 2
    c) 1
    d) 1/2
    e) 1/4

    Answer: e Difficulty: 2

48. An astronomical telescope operates at f/12 and has an aperture of 36.cm. What is its angular magnification when used with a 2.0 cm eyepiece?
    a) 210
    b) 18
    c) 6
    d) 72
    e) 144

    Answer: a Difficulty: 2

49. Christie bought a 9 X 50 pair of binoculars for stellar observing. The 9 is the:
    a) aperture in cm
    b) aperture in mm
    c) magnification
    d) f/number

    Answer: c Difficulty: 2

50. Jennifer bought a 7 X 25 pair of binoculars for observing baseball. The 25 describes the:
    a) aperture in cm
    b) aperture in mm
    c) magnification
    d) f/number

    Answer: b Difficulty: 2

ESSAY

51. Many Earth-based telescopes are larger than the Hubble Space Telescope, so what justifies the great expense of a "space-based" instrument?

    Answer:
    The atmosphere distorts images and usually limit resolution. A telescope in the vacuum of space can achieve its intrinsic resolution and yield pictures of with relatively high detail. It also can use parts of the spectrum which get absorbed by the atmosphere.

    Difficulty: 2

FILL-IN-THE-BLANK

52. The 94 inch diameter Hubble Space Telescope has what intrinsic angular resolution at 400.nm?

    Answer: 0.20 microradians = 0.042 arc-seconds   Difficulty: 2

ESSAY

53. Both the microscope and the telescope make the object appear much larger. Why are these two instruments designed differently (e.g. why are the equations for magnification quite different)?

    Answer:
    The object for a telescope is at infinity whereas the object of a microscope is very close to the objective. The ray diagrams and the designs are thus quite different.
    Difficulty: 2

FILL-IN-THE-BLANK

54. Mary builds a simple magnifying telescope using two lenses of focal length 80.cm and 5.0 cm. The respective diameters are 60.mm and 10.mm. How many degrees wide does the moon appear through the telescope if it appears only 1/2 degree wide with the "naked eye"?

    Answer: 8 degrees wide (16 times larger)   Difficulty: 2

MULTIPLE CHOICE

55. Pat is viewing the moon with her new telescope. She is using a 10.mm eyepiece. Which of the following eyepieces will double the magnification?
    a) 40.Diopter
    b) 12.mm
    c) 200.Diopter
    d) 8.mm
    e) 5.Diopter
    f) 100.Diopter

    Answer: c   Difficulty: 2

Chapter 25

56. Which of the following binoculars would be best for observing very faint comets among the stars?
   a) 7 X 60
   b) 12 X 50
   c) 10 X 35
   d) 8 X 50
   e) 5 X 20

   Answer: a   Difficulty: 2

FILL-IN-THE-BLANK

57. What is the minimum distance of two points on the moon that can be just resolved by the Keck 10.meter telescope in Hawaii (use a wavelength of 600.nm, lunar distance of 380.thousand Km)
   a) if there were no atmosphere?
   b) if the atmosphere limits resolution to 0.2 arc-sec?

   Answer: a) 28.meters   b) 0.37 Km   Difficulty: 2

MULTIPLE CHOICE

58. Assuming a pupil diameter of 3.mm, what is the angular resolution of the human eye at the middle of the visible spectrum (550.nm)?
   a) 0.8 degree
   b) 0.8 arc-minute
   c) 0.2 degree
   d) 0.2 arc-minute
   e) 2. milliradian
   f) 0.08 radian

   Answer: b

59. Mixing pigments is an example of the _____ method of color production.
   a) additive
   b) subtractive
   c) complementary
   d) composite

   Answer: b   Difficulty: 1

60. When the cornea or lens is out of round, the eye suffers from:
   a) myopia
   b) presbyteria
   c) hyperopia
   d) astigmatism
   e) blindness

   Answer: d   Difficulty: 2

ESSAY

61. How is it possible for the angular magnification to be large and, at the same time, for the lateral magnification to be small. Give an example.

    Answer:
    The angular mag. is the ratio of ANGULAR size looking through the instrument compared to the angular size without. The lateral mag. is the ratio of the actual size of image compared to that of the object. The two magnifications are comparing different things. As an example: the moon may appear larger (in angle) through a telescope but the image is certainly smaller than the moon itself.

    Difficulty: 2

62. Explain how the eye and a camera differ in adjusting for a change in focus.

    Answer:
    The camera lens has a fixed focal length so it adjusts focus by varying the lens-film distance. The eye has an essentially fixed lens-retina distance so the focal length of the lens is varied by changing the lens curvature (by muscular action).

    Difficulty: 2

MULTIPLE CHOICE

63. Kelli examines her new ruby ring with the jeweler's "loop". The ruby appears 7.0 times larger. If her near point is 35.cm, what is the focal length of the lens?
    a) 36.mm
    b) 42.mm
    c) 50.mm
    d) 70.mm

    Answer: c

Chapter 26

MATCHING

Match the unit to the physical quantity.

a) m-°K
b) J-s
c) J
d) m
e) m^{-1}

1. Wien constant

   Answer: a) m-°K Difficulty: 2

2. Planck's constant

   Answer: b) J-s Difficulty: 2

3. work function

   Answer: c) J Difficulty: 2

4. Compton wavelength

   Answer: d) m Difficulty: 2

5. Rydberg constant

   Answer: e) m^{-1} Difficulty: 2

MULTIPLE CHOICE

6. A blackbody is an ideal system that
   a) absorbs 100% of the light incident upon it, but cannot emit light of its own (i.e., a "black" body).
   b) emits 100% of the light it generates, but cannot absorb radiation of its own.
   c) either absorbs 100% of the light incident upon it, or emits 100% of the radiation it generates.
   d) absorbs 50% of the light incident upon it, and emits 50% of the radiation it generates.

   Answer: c Difficulty: 2

7. The ultraviolet catastrophe is a catastrophe because
   a) human vision cannot detect ultraviolet light directly.
   b) Wien's law predicts an infinite intensity of radiation at small wavelengths.
   c) Wien's law predicts an infinite intensity of radiation at large wavelengths.
   d) Ultraviolet light will not produce electrons by the photoelectric effect.
   e) Classical physics predicts that an infinite number of electrons will be produced by ultraviolet light when incident upon a photoelectric material.

   Answer: b   Difficulty: 2

8. What is a photon?
   a) An electron in an excited state
   b) A small packet of electromagnetic energy that has particle-like properties
   c) One form of a nucleon, one of the particles that makes up the nucleus
   d) An electron that has been made electrically neutral
   e) Another name for a neutrino

   Answer: b   Difficulty: 2

9. A photon is a particle that
   a) has zero electric charge.
   b) has zero electric field associated with it.
   c) cannot travel in a vacuum.
   d) has a velocity in a vacuum that varies with the photon frequency.
   e) is usually observed only in connection with nuclear reactions.

   Answer: a   Difficulty: 2

10. Which of the following is an accurate statement?
    a) In vacuum, ultraviolet photons travel faster than infrared photons.
    b) Photons can have positive or negative charge.
    c) Infrared photons have enough energy to readily break bonds in a nucleic acid molecule like DNA.
    d) An ultraviolet photon has more energy than does an infrared photon.
    e) Photons do not have momentum (i.e., they cannot exert pressure on things).

    Answer: d   Difficulty: 2

ESSAY

11. What is the energy, in eV, of a photon of wavelength 550 nm?

    Answer: 2.26 eV   Difficulty: 2

Chapter 26

12. What is the wavelength of a photon of energy 2.0 eV?

    Answer: 622 nm Difficulty: 2

13. A small gas laser of the type used in classrooms may radiate light at a power level of 2 mW. If the wavelength of the emitted light is 642 nm, how many photons are emitted per second?

    Answer: 6.5×10^{14} Difficulty: 2

14. The human eye can just detect green light of wavelength 500 nm, which arrives at the retina at the rate of 2×10^{-18} W. How many photons arrive each second?

    Answer: 5 Difficulty: 2

MULTIPLE CHOICE

15. A beam of red light and a beam of violet light each deliver the same power on a surface. For which beam is the number of photons hitting the surface per second the greatest?
    a) The red beam
    b) The violet beam
    c) The number of photons per second is the same for both beams.
    d) This cannot be answered without knowing just what the light intensity is.
    e) This cannot be answered without knowing just what the wavelengths of the light used are.

    Answer: a Difficulty: 2

16. Which color of light has the lowest energy photons?
    a) Red
    b) Yellow
    c) Green
    d) Blue

    Answer: a Difficulty: 2

17. The energy of a photon depends on
    a) its amplitude.
    b) its velocity.
    c) its frequency.
    d) none of the above.

    Answer: c Difficulty: 2

TRUE/FALSE

18. You've just built for yourself a solar-powered toy car, which moves when you shine a flashlight on its solar cells. If you shine another flashlight, which is brighter than the first, on it, your car will go faster.

    Answer: F Difficulty: 2

MULTIPLE CHOICE

19. The ratio of energy to frequency for a given photon gives
    a) its amplitude.
    b) its velocity.
    c) Planck's constant.
    d) its work function.

    Answer: c Difficulty: 2

20. How much energy, in joules, is carried by a photon of wavelength 660 nm?
    a) 1.46×10^{-48} J
    b) $3.01 \times 10^{-1} \mu$ J
    c) 6.63×10^{-34} J
    d) None of the above

    Answer: b Difficulty: 2

21. When the accelerating voltage in an X-ray tube is doubled, the minimum wavelength of the X-rays
    a) is increased to twice the original value.
    b) is increased to four times the original value.
    c) is decreased to one-half the original value.
    d) is decreased to one-fourth the original value.

    Answer: c Difficulty: 2

ESSAY

22. What are the shortest wavelength X-rays produced by a 50,000-V X-ray tube?

    Answer: 2.49×10^{-11} m Difficulty: 2

23. A metallic surface has a work function of 2.5 eV. What is the longest wavelength that will eject electrons from the surface of this metal?

    Answer: 497 nm Difficulty: 2

Chapter 26

MULTIPLE CHOICE

24. Planck's constant
   a) sets an upper limit to the amount of energy that can be absorbed or emitted.
   b) sets a lower limit to the amount of energy that can be absorbed or emitted.
   c) relates mass to energy.
   d) does none of the above.

   Answer: b Difficulty: 2

ESSAY

25. A metallic surface is illuminated with light of wavelength 400 nm. If the work function for this metal is 2.4 eV, calculate the kinetic energy of the ejected electrons, in electron-volts.

   Answer: 0.604 eV Difficulty: 2

26. At what rate are photons emitted by a 50-W sodium vapor lamp? (Assume that the lamp's light is monochromatic and of wavelength 589 nm.)

   Answer: 1.48×10^{20} per second Difficulty: 2

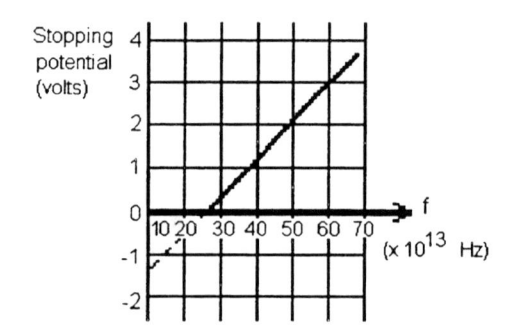

27. The graph shown is a plot based on student data from their testing of a photoelectric material. Determine (a) the cutoff requency. (b) the work function.

   Answer: (a) 25×10^{13} Hz   (b) $1.7 \times 10^{-1} \mu$ J Difficulty: 2

28. Molybdenum has a work function of 4.2 eV. What is the kinetic energy of the electrons emitted in the photoelectric effect when illuminated with light of wavelength 366 nm?

   Answer: 0.8 eV Difficulty: 2

MULTIPLE CHOICE

29. A metal surface is illuminated with blue light and electrons are ejected at a given rate each with a certain amount of energy. If the intensity of the blue light is increased, electrons are ejected
    a) at the same rate, but with more energy per electron.
    b) at the same rate, but with less energy per electron.
    c) at an increased rate with no change in energy per electron.
    d) at a reduced rate with no change in energy per electron.

    Answer: c Difficulty: 2

30. In the photoelectric effect, the energies of the ejected electrons
    a) are proportional to the speed of light.
    b) are proportional to the intensity of light.
    c) are proportional to the frequency of light.
    d) vary randomly.

    Answer: c Difficulty: 2

31. In order for a photon to eject an electron from a metal's surface in the photoelectric effect, the photon's
    a) frequency must be greater than a certain minimum value.
    b) speed must be greater than a certain minimum value.
    c) wavelength must be greater than a certain minimum value.
    d) momentum must be zero.

    Answer: a Difficulty: 2

32. The photoelectric effect is explainable assuming
    a) that light has a wave nature.
    b) that light has a particle nature.
    c) that light has a wave nature and a particle nature.
    d) none of the above.

    Answer: b Difficulty: 2

ESSAY

33. X-rays with a wavelength of 0.00100 nm are scattered by free electrons at 130°. What is the kinetic energy of each recoil electron?

    Answer: 992 keV Difficulty: 2

Chapter 26

MULTIPLE CHOICE

34. In a Compton scattering experiment, what scattering angle produces the greatest change in wavelength?
    a) Zero degrees
    b) 90°
    c) 180°
    d) None of the above

    Answer: c Difficulty: 2

35. In the Compton effect, as the scattering angle increases, the frequency of the X-rays scattered at that angle
    a) increases.
    b) decreases.
    c) does not change.
    d) varies randomly.

    Answer: b Difficulty: 2

36. What change in wavelength is expected when X-rays are scattered at 90°?
    a) No change at this angle
    b) 4.84×10^{-3} nm
    c) 2.42×10^{-3} nm
    d) 1.18×10^{-3} nm

    Answer: c Difficulty: 2

37. The energy required to ionize a hydrogen atom is
    a) 27.2 eV
    b) 13.6 eV
    c) 6.8 eV
    d) 3.4 eV

    Answer: b Difficulty: 2

ESSAY

38. What is the radius of the electron orbit in a singly ionized helium ion?

    Answer: 0.027 nm Difficulty: 2

39. What is the energy, in eV, of the n = 3 state in hydrogen?

    Answer: -1.5 eV Difficulty: 2

40. What is the ionization energy for a doubly ionized lithium ion (Z = 3)?

    Answer: 122 eV Difficulty: 2

Testbank

MULTIPLE CHOICE

41. When an electron jumps from an orbit where n = 4 to one where n = 2
    a) a photon is emitted.
    b) a photon is absorbed.
    c) two photons are emitted.
    d) two photons are absorbed.
    e) none of the above occur.

    Answer: a Difficulty: 2

42. The distance between adjacent orbit radii in a hydrogen atom
    a) increases with increasing values of n.
    b) decreases with increasing values of n.
    c) remains constant for all values of n.
    d) varies randomly with increasing values of n.

    Answer: a Difficulty: 2

43. The energy difference between adjacent orbit radii in a hydrogen atom
    a) increases with increasing values of n.
    b) decreases with increasing values of n.
    c) remains constant for all values of n.
    d) varies randomly with increasing values of n.

    Answer: b Difficulty: 2

44. In state n = 1, the energy of the hydrogen atom is -13.58 eV. What is its energy in state n = 2?
    a) -6.79 eV
    b) -4.53 eV
    c) -3.40 eV
    d) -1.51 eV

    Answer: c Difficulty: 2

45. In making a transition from state n = 1 to state n = 2, the hydrogen atom must
    a) absorb a photon of energy 10.2 eV.
    b) emit a photon of energy 10.2 eV.
    c) absorb a photon of energy 13.58 eV.
    d) emit a photon of energy 13.58 eV.

    Answer: a Difficulty: 2

46. What is the ionization energy of the neutral hydrogen atom?
    a) 27.2 eV
    b) 13.6 eV
    c) 6.8 eV
    d) None of the above

    Answer: b Difficulty: 2

Chapter 26

47. What is the ionization energy of singly ionized helium?
   a) 54.4 eV
   b) 27.2 eV
   c) 13.6 eV
   d) None of the above

   Answer: a Difficulty: 2

48. Consider an atom with four accessible energy levels. What is the maximum number of different wavelengths that could be emitted by such an atom?
   a) 4
   b) 5
   c) 6
   d) 7

   Answer: c Difficulty: 2

49. In order for the atoms in a neon discharge tube (or a neon beer sign) to emit their characteristic orange light, it is necessary that
   a) each atom carry a net electric charge.
   b) the atoms be continually replaced with fresh atoms, because the energy of the atoms tends to be used up with continued excitation, resulting in dimmer and dimmer light.
   c) there be no unoccupied energy levels in each atom.
   d) the neon nucleus be unstable.
   e) electrons first be given energy to raise them from their ground state to an excited state.

   Answer: e Difficulty: 2

50. In a famous experiment done at the end of the 19th century, two metal electrodes were placed in an evacuated glass tube. A high voltage was applied between them. By using an appropriate metal for the cathode, (e.g. sodium or potassium), it was found that a current could be made to flow when the cathode was illuminated with blue light, but no current would flow when red light was used, no matter how bright the light. Einstein was able to explain this strange phenomenon, and as a result, this experiment was considered the first experimental demonstration of
   a) the particle nature of light.
   b) the wave nature of electrons.
   c) the wave nature of light.
   d) the particle nature of electrons.
   e) the equivalency of electron and photons.

   Answer: a Difficulty: 2

51. You may have heard it said that all objects, even ones at the temperature of our bodies, are continually emitting electromagnetic radiation. Is this true?
    a) Yes, and the radiation referred to results from the trace amounts of radioactive material that have accumulated in our bodies from the food we eat.
    b) Yes, and the radiation from our bodies is mostly in the infrared region of the spectrum, and hence not detected by our eyes.
    c) Yes, and this radiation is mostly in the form of radio waves, which is a common source of static often heard on an AM radio.
    d) No, this is not true. If a person were emitting electromagnetic radiation, he could be seen glowing in the dark, and this is obviously not the case for most people.
    e) No, this is not true, because our bodies are electrically neutral.

    Answer: b Difficulty: 2

TRUE/FALSE

52. Bound electrons have negative energy.

    Answer: T Difficulty: 2

53. Photons are emitted when a bound electron jumps up to a higher energy state.

    Answer: F Difficulty: 2

MULTIPLE CHOICE

54. The word LASER is an acronym for
    a) Light Altered Spectra of Energy Radiated.
    b) LAtent Source of Enhanced Radiation.
    c) Light Amplification by the Stimulated Emission of Radiation.
    d) Light Absorbed States of Energetic Resonance.

    Answer: c Difficulty: 2

55. In order to produce a hologram, one needs, in addition to an object and a piece of photographic film,
    a) a beam of monochromatic light and a mirror.
    b) a beam of monochromatic light and a lens.
    c) a beam of coherent light and a lens.
    d) a beam of coherent light and a mirror.

    Answer: d Difficulty: 2

Chapter 26

56. When a hologram is illuminated with a beam of coherent light, it produces
   a) both a real and a virtual image.
   b) only a real image of the object.
   c) only a virtual image of the object.
   d) none of the above.

   Answer: a Difficulty: 2

TRUE/FALSE

57. A hologram shows a magnifying glass held over a postage stamp. You accidentally drop the hologram, and it shatters into many small pieces. The magnifier and stamp will be 100% visible in any fragment you look at.

   Answer: T Difficulty: 2

MULTIPLE CHOICE

58. How much energy would be required to accelerate a 77.Kg professor from rest to 90% the speed of light?
   a) 0.30 mc^2
   b) 0.41 mc^2
   c) 1.3 mc^2
   d) 2.3 mc^2

   Answer: c Difficulty: 2

FILL-IN-THE-BLANK

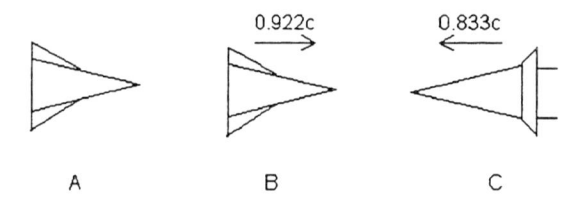

59. A person onboard spaceship A sees craft B moving away at 92.2% the speed of light and he sees C approaching at 0.833c (see figure).
   A) Spacecraft B sees C approaching with what speed?
   B) Ship C sees B approaching with what speed?
   C) Ship C sees A approaching at what speed?
   D) Spacecraft B determines A to be approaching at what speed?

   Answer:
   A) approaches at 0.993c   B) approaches at 0.933c
   C) approaches at 0.833c   D) approaches at 0.922c

   Difficulty: 2

MULTIPLE CHOICE

60. Albert Michelson is primarily remembered for:
    a) proving length contraction
    b) not being able to detect an "ether"
    c) his theory published in 1916
    d) proving that the speed of light is not constant

    Answer: b Difficulty: 1

61. Compared to "Special" Relativity, "General" Relativity is more concerned with:
    a) gravitation
    b) mass-energy
    c) time dilation
    d) Unified fields
    e) Lorentz transformations

    Answer: a Difficulty: 1

62. According to the special theory of relativity, if you were in a space ship moving at 87% the speed of light, you would notice your pulse rate to be:
    a) twice as fast
    b) half as fast
    c) 4 times as much
    d) 4 times slower
    e) unchanged

    Answer: e Difficulty: 1

FILL-IN-THE-BLANK

63. State the two fundamental postulates of Einstein's Special Relativity.

    Answer:
    1) the constancy of the speed of light, 2) the laws of nature are the same in all inertial frames.
    Difficulty: 2

Chapter 26

MULTIPLE CHOICE

64. When star light passes by the sun:
   a) it is unaffected by the sun
   b) it is deflected away from the sun
   c) it is deflected toward the sun
   d) it is absorbed by the sun

   Answer: c Difficulty: 1

65. At what speed would a 100.meter long spaceship appear to be 60.meters long ?
   a) 24. Mm/s
   b) 2.4 Gm/s
   c) 240.Mm/s
   d) 2400.Km/s
   e) 0.24 m/s
   f) 2400.Mm/s

   Answer: c Difficulty: 2

66. In spite of the fictional exploits of the starship Enterprise, "Relativity" teaches us that the maximum possible speed is:
   a) "warp" 0
   b) "warp" 1
   c) "warp" 10
   d) "warp" infinity

   Answer: b Difficulty: 2

FILL-IN-THE-BLANK

67. A muon at rest decays in 2.2 μs. Moving at 99.% the speed of light, it would be seen to "live" for how long?

   Answer: 16.μs. Difficulty: 2

68. A starship is built to plans that state it is to be constructed 100.meters long. It is launched and as it coasts by the Earth, it is seen to be shortened to 5.48 meters long.
   A) How fast is it moving?
   B) The ship's crew measure the length of the ship to be how long?

   Answer: A) 99.85% the speed of light B) 100.meters Difficulty: 2

69. If the mass of a 1.0 kg book could be entirely converted into elctrical energy, how many Kw-h would that generate?

   Answer: 2.5 x 10^{10} Kw-h (= 9.0 x 10^{16} J) Difficulty: 2

70. Above what speed do the Relativistic and the Newtonian expressions for KINETIC ENERGY disagree by more than 1%?

    Answer: v/c = 0.12 (v= 3.6x10^7 m/s) Difficulty: 2

ESSAY

71. Using the Equivalence Principle as a guide, explain how one might cancel out (or make disappear) a gravitational field.

    Answer:
    If the observer accelerates in the direction of the gravitational field with the "acceleration of gravity" then all local gravitational effects will apparently disappear.
    Difficulty: 2

72. Using the velocity transformation formula, calculate the speed in the "rest frame" of a particle which is seen to be moving at c in a "moving frame".

    Answer:
    u = (v+u')/(1+vu'/c^2) = (v+c)/(1+vc/c^2) = c(v+c)/(c+v) = c   Light speed in one frame will be light speed in all inertial frames.
    Difficulty: 2

FILL-IN-THE-BLANK

73. Consider a student whose mass is 70.kg (he weighs 154.lb). If all his mass were converted to energy, this energy could light a 100.watt lamp for how long?

    Answer: 6.3 x 10^{10} s = 2.0 billion years! Difficulty: 2

MULTIPLE CHOICE

74. A spaceship, traveling at WARP 1/2 away from the Earth, launches a shuttle moving away from the ship at c/2 (c/2 seen from the ship). Classically one expects to see the shuttle moving at c from the Earth, but relativistically what speed do we expect seen from the Earth?
    a) 0.5c
    b) 0.6c
    c) 0.7c
    d) 0.8c
    e) 0.9c
    f) 1.0c

    Answer: d Difficulty: 2

Chapter 26

FILL-IN-THE-BLANK

75. From the Earth, they see the Enterprise approaching at 0.800c and the Klingons approaching from the opposite direction at 0.900c. From the Enterprise, the crew sees the Klingon ship approaching at what speed?

    Answer: Klingons approaching the Enterprise at 0.988c Difficulty: 2

76. Suppose a 70.kg student were to become (be compressed) a blackhole.
    a) What would be his/her Schwarzshild radius ("size" of the B.H.)?
    b) How does this compare to the size of a proton?

    Answer:
    a) 1.0×10^{-25} meters
    b) About 10 orders of magnitude smaller than a proton.
    Difficulty: 2

MULTIPLE CHOICE

77. If one uses the classical expression for kinetic energy ($1/2\ m\ v^2$) for a particle traveling at half the speed of light, the result will be deficient by how many percent?
    a) 1%
    b) 7%
    c) 13%
    d) 19%
    e) 27%
    f) 51%

    Answer: d Difficulty: 2

78. A stick is moving along the x axis at 90%c. A person riding with the stck sees it inclined at 45.° with respect to the x axis. Observers in the rest frame see the stick making what angle with respect to the x axis?
    a) 45.°
    b) 66.°
    c) 79.°
    d) 88.°

    Answer: b Difficulty: 2

Chapter 27

MATCHING

Match the description to the physicist.

a) developed a wave equation for matter waves (1926)
b) suggested the existence of matter waves (1924)
c) first to produce X-ray-like diffraction patterns of electrons passing through thin metal foil
d) predicted the positron from relativistic quantum mechanics (1928)
e) discovered the positron using a cloud chamber (1932)
f) first to produce diffraction patterns of electrons in crystals (1927)
g) set the limits on the probability of measurement accuracy (1927)

1. Erwin Schrodinger

    Answer: a) developed a wave equation for matter waves (1926)

    Difficulty: 2

2. Louis de Broglie

    Answer: b) suggested the existence of matter waves (1924) Difficulty: 2

3. G. P. Thomson

    Answer:
    c) first to produce X-ray-like diffraction patterns of electrons passing through thin metal foil
    Difficulty: 2

4. Paul Dirac

    Answer: d) predicted the positron from relativistic quantum mechanics (1928)

    Difficulty: 2

5. C. D. Anderson

    Answer: e) discovered the positron using a cloud chamber (1932)

    Difficulty: 2

6. C. J. Davisson & L. H. Germer

    Answer:
    f) first to produce diffraction patterns of electrons in crystals (1927)
    Difficulty: 2

Chapter 27

7. Werner Heisenberg

   Answer: g) set the limits on the probability of measurement accuracy (1927)

   Difficulty: 2

ESSAY

8. Atoms in crystals are typically separated by distances of 0.1 nm. What KE must an electron have, in eV, in order to have a wavelength of 0.1 nm?

   Answer: 151 eV Difficulty: 2

9. The radius of a typical nucleus is about 5×10^{-15} m. Assuming this to be the uncertainty in the position of a proton in the nucleus, estimate the uncertainty in the proton's energy (in eV).

   Answer: 412 keV Difficulty: 2

MULTIPLE CHOICE

10. As a particle travels faster, its de Broglie wavelength
    a) increases.
    b) decreases.
    c) remains constant.
    d) could be any of the above; it depends on other factors.

    Answer: b Difficulty: 2

11. When a photon is scattered from an electron, there will be an increase in its
    a) energy.
    b) frequency.
    c) wavelength.
    d) momentum.

    Answer: b Difficulty: 2

TRUE/FALSE

12. Ordinary photographic film is least sensitive to red light.

    Answer: T Difficulty: 2

MULTIPLE CHOICE

13. When an electron is accelerated to a higher speed, there is a decrease in its
    a) energy.
    b) frequency.
    c) wavelength.
    d) momentum.

    Answer: c Difficulty: 2

14. The reason the wavelike nature of a moving baseball is not noticed in everyday life is that
    a) it doesn't have a wavelike nature.
    b) its wavelength is too small.
    c) its frequency is too small.
    d) its energy is too small.
    e) no one pays attention to such things except for the Mets; and they can't hit a curve ball anyway.

    Answer: b Difficulty: 2

15. The part of an electron microscope that plays the same role as the lenses do in an optical microscope is
    a) the vacuum chamber.
    b) the coils.
    c) the cathode.
    d) the deflector plates.

    Answer: b Difficulty: 2

16. Which of the following microscopes is capable of "photographing" individual atoms?
    a) Light microscope
    b) Scanning tunneling microscope
    c) Transmission electron microscope
    d) Scanning electron microscope

    Answer: b Difficulty: 2

17. What is the de Broglie wavelength of a ball of mass 200 g moving with a speed of 30 m/s?
    a) 1.1×10^{-34} m
    b) 2.2×10^{-34} m
    c) 4.5×10^{-28} m
    d) 6.67×10^{-27} m

    Answer: a Difficulty: 2

Chapter 27

18. An electron has a wavelength of 0.123 nm. What is its energy in eV? (This energy is not in the relativistic region.)
    a) 20 eV
    b) 60 eV
    c) 80 eV
    d) 100 eV

    Answer: d Difficulty: 2

19. A person of mass 50 kg has a wavelength of 4.42 x 10^{-36} m when running. How fast is she running?
    a) 2 m/s
    b) 3 m/s
    c) 4 m/s
    d) 5 m/s

    Answer: b Difficulty: 2

20. Although de Broglie had not yet made his discovery of the wavelength associated with electrons at the time Bohr advanced his model of the atom, one can see how such an idea fits in with Bohr's ideas concerning electron orbits. The basic idea is that
    a) since electrons are waves, they can be anywhere in the atom.
    b) the wave associated with an electron must interfere constructively with itself, which suggests an orbit circumference should be a multiple of the electron wavelength.
    c) the diameter of the orbit will be equal to the electron wavelength.
    d) since electrons are waves, their frequency of revolution will be given by v/λ, where v is the electron velocity in an orbit.
    e) electrons would be expected to emit light of well-defined sharp wavelengths, since they themselves have definite wavelengths.

    Answer: b Difficulty: 2

21. What advantage might an electron microscope have over a light microscope?
    a) Electrons are more powerful.
    b) Shorter wavelengths are possible.
    c) Longer wavelengths are possible.
    d) None.

    Answer: b Difficulty: 2

ESSAY

22. Suppose that a particle is confined in a one-dimensional box that extends from the origin to x = a. If the wave function of the particle is given by Ax, where A is a constant, what is the probability of finding the particle in the right half of the box?

    Answer: 75% Difficulty: 2

23. Suppose that a particle is confined to a one-dimensional box that extends from x = -L to x = +L. Its wave function is 4(x) = A cos(5 π x/2L). What is the probability of finding the particle in the range from x = L/5 to x = L?

    Answer: 40% Difficulty: 2

24. If the energy of a particle in a box in the ground state is E, what is the energy in the state n = 4?

    Answer: 4E Difficulty: 2

25. An electron is confined to a one-dimensional box of width 0.1 nm, about the size of an atom. What is its energy, in eV, in the ground state?

    Answer: 38 eV Difficulty: 2

MULTIPLE CHOICE

26. Suppose that a particle is trapped in a box. How does the spacing in energy between adjacent energy levels change as the quantum number identifying a level increases?
    a) The spacing remains constant.
    b) The spacing increases.
    c) The spacing decreases.
    d) The spacing alternately increases and decreases, depending on whether the wave function is a sine or a cosine function.

    Answer: b Difficulty: 2

27. What effect does the size of the box have on the ground state energy of a particle trapped in a box?
    a) If the box is smaller, the energy is greater.
    b) If the box is larger, the energy is greater.
    c) The size of the box has no influence on the energy.
    d) Not enough information is given to answer this question.

    Answer: a Difficulty: 2

ESSAY

28. An electron is known to be confined to a region of width 0.1 nm. What is an approximate expression for the least kinetic energy it could have, expressed in eV?

    Answer: 3.8 eV Difficulty: 2

29. An electron is trapped in a box of width 0.1 nm. What is the wavelength of the photon emitted when an electron makes a transition from the n = 3 to the n = 1 state?

    Answer: 11 nm Difficulty: 2

Chapter 27

30. Suppose that the speed of an electron traveling 2000 m/s is known to an accuracy of 1 part in 10^5 (i.e., within 0.01%). What is the greatest possible accuracy within which we can determine the position of this electron?

    Answer: 5.8 mm Difficulty: 2

MULTIPLE CHOICE

31. Heisenberg's Uncertainty Principle states that
    a) all measurements are to some extent inaccurate, no matter how good the instrument used.
    b) we cannot in principle know simultaneously the position and momentum of a particle with absolute certainty.
    c) we can never be sure whether a particle is a wave or a particle.
    d) at times an electron appears to be a particle and at other times it appears to be a photon.
    e) the charge on the electron can never be known with absolute accuracy.

    Answer: b Difficulty: 2

32. Neutrons of energy 0.025 eV pass through two slits 1.0 mm apart. How far apart will peaks in beam intensity (due to constructive interference) be on a screen 0.5 m away?
    a) 90 nm
    b) 105 nm
    c) 150 nm
    d) 200 nm

    Answer: a Difficulty: 2

33. The uncertainty in the position of a proton is 0.053 nm. What is the uncertainty in its speed?
    a) 1200 m/s
    b) 975 m/s
    c) 600 m/s
    d) 2.2 m/s

    Answer: a Difficulty: 2

34. Suppose you were to try to create a proton-antiproton pair by annihilation of a very high energy photon. The proton and the anti-proton have the same masses, but opposite charges. What energy photon would be required?
    a) 1.022 MeV
    b) 1880 MeV
    c) 940 MeV
    d) 223 MeV

    Answer: b Difficulty: 2

35. A positively charged electron is
    a) called a proton.
    b) called a positron.
    c) also called an electron.
    d) not found in nature.

    Answer: b Difficulty: 2

36. One reason a photon could not create an odd number of electrons and positrons is that such a process would
    a) not conserve charge.
    b) not conserve energy.
    c) require photon energies that are not attainable.
    d) result in the creation of mass.

    Answer: a Difficulty: 2

37. When a positronium atom decays, it emits two photons. What are the energies of the photons?
    a) 0.511 MeV and 1.022 MeV
    b) 0.225 MeV and 0.285 MeV
    c) 0.256 MeV and 0.256 MeV
    d) 0.511 MeV and 0.511 MeV

    Answer: d Difficulty: 2

38. What is created in pair production?
    a) A proton and an electron
    b) A proton and an antiproton
    c) An electron and a positron
    d) An electron and a photon

    Answer: c Difficulty: 2

39. A photon cannot create an electron pair unless its frequency is:
    a) less than 1.24×10^{20} Hz
    b) less than 2.47×10^{20} Hz
    c) greater than 1.24×10^{20} Hz
    d) greater than 2.47×10^{20} Hz

    Answer: c Difficulty: 2

ESSAY

40. In the Bohr Theory of the hydrogen atom, what very non-classical assumption was made by Bohr?

    Answer: He assumed angular momentum was quantized: $L = n\, h/(2\pi)$

    Difficulty: 2

Chapter 27

FILL-IN-THE-BLANK

41. Match the following traditional symbols
    ___Speed of light   (A) μ_o
    ___Boltzmann's constant   (B) e
    ___Electron charge   (C) ϵ_o
    ___Avogadro's number   (D) N_A
    ___permeability of free space (E) c
    ___permittivity of free space (F) k

Answer: E F B D A C   Difficulty: 2

MULTIPLE CHOICE

42. In the Bohr theory the orbital radius depends upon the principle quantum number in what way?
    a) n
    b) 1/n
    c) n^2
    d) $1/n^2$
    e) n^3
    f) $1/n^3$
    g) ln(n)

Answer: c Difficulty: 2

ESSAY

43. How would Bohr's theory be changed if only gravitation held the electron to the nucleus?

Answer:
The force would be GmM/r^2 instead of ke^2/r^2 so replace ke^2 with GmM in all Bohr results. e.g. $E_n = 2\pi m(GmM)^2/(hn)^2$
Difficulty: 2

FILL-IN-THE-BLANK

44. Mathilda collects photoelectric data in her physics laboratory. She measures a stopping potential of 5.82 volts for radiation of wavelength 100.nm, and 17.99 volts stops 50.0 nm.
A) What is the work function for the material in eV?
B) From her data one determines Planck's constant to be what?

Answer: A) 6.35 eV   B) 6.50 x 10^{-34} J-s   Difficulty: 2

Testbank

TRUE/FALSE

45. The compact disc stores information by means of holography.

    Answer: F Difficulty: 1

FILL-IN-THE-BLANK

46. Is the maximum kinetic energy of a photo-electron emitted from a metal in the photoelectric effect proportional to the frquency of the incident light?

    Answer:
    Not unless the work function is zero. K=hv-ϕ so (K+ϕ) is proproportional to frequency.
    Difficulty: 2

47. If in Compton scattering of X-rays from a metal, the wavelength increases by 5% at an angle of 60° ; what was the original wavelength before scattering?

    Answer: 2.4×10^{-11} m Difficulty: 2

TRUE/FALSE

48. Classical theory predicted that the photocurrent, of the photoelectric effect, should be proportional to the intensity of the light.

    Answer: T Difficulty: 1

FILL-IN-THE-BLANK

49. The cosmic background radiation has been measured to have a thermodynamic temperature of 2.7° K.
    a) at what wavelength does the radiation peak?
    b) to what part of the electromagnetic spectrum does this correspond?

    Answer: a) λ = 1.1 mm   b) microwave Difficulty: 2

MULTIPLE CHOICE

50. The fact that $\lambda_m = 2.9 \times 10^{-3}$ m-K is known as:
    a) Planck's law
    b) Wien's law
    c) Bohr's law
    d) de Broglie's law
    e) Sommerfield's law
    f) Maxwell's law

    Answer: b Difficulty: 2

Chapter 27

51. If sunlight peaks at 0.55 μm wavelength, this infers a surface temperature of the Sun of how many thousand degrees Kelvin?
    a) 2.5
    b) 5.5
    c) 9.5
    d) 15.
    e) 25.

    Answer: b Difficulty: 2

52. Albert Einstein received the 1921 Nobel Prize for his:
    a) Special Thoery of Relativity
    b) General Theory of Relativity
    c) Theory of the Photoelectric Effect
    d) Theory of the Brownian Motion

    Answer: c Difficulty: 2

TRUE/FALSE

53. Above the CUTOFF FREQUENCY, photons do not have enough energy to dislodge photoelectrons and no photocurrent is observed.

    Answer: F Difficulty: 2

ESSAY

54. What was it that the Compton Effect illustrated?

    Answer:
    Electromagnetic waves (X-rays) could act as particles (photons) when scattering off electrons.
    Difficulty: 2

55. Classical theory predicts that the planets can orbit the Sun quite stably. Why could not stable orbits be predicted for the atom by simply substituting the electrical attraction for the gravitational attraction? Why was there a classical crisis in describing the atom?

    Answer:
    An electron, classically orbiting the nucleus, is accelerating and electromagnetic theory says it should radiate and thus the orbit would rapidly decay. A long lived stable atom could not be explained classically.
    Difficulty: 2

56. Bohr's atomic theory predicted energy levels for hydrogen of $E_n = -(2\pi^2 k^2 e^4 m)/(hn)^2$. If instead we place 1 electron in "orbit" about a bare nucleus of atomic number Z, How do you expect Z to be incorporated into the E_n formula and why?

Answer:
The attractive force is $k(Ze)e/r^2$ instead of ke^2/r^2. So replace e^2 by Ze^2 in the Bohr predictions:
$E_n = -(2\pi^2 k^2 Z^2 e^4 m)/(hn)^2$

Difficulty: 2

FILL-IN-THE-BLANK

57. Give an example of a material whose useful property depends upon a metastable state.

Answer: Phosphorescent materials. Difficulty: 2

MULTIPLE CHOICE

58. Light from a gas discharge tube gives a (an) _____ spectrum.
    a) continuous
    b) emission
    c) absorption
    d) blackbody

Answer: b Difficulty: 2

59. Which hydrogen spectral series falls in the visible region?
    a) Paschen
    b) Lyman
    c) Balmer
    d) Brackett

Answer: c Difficulty: 2

Chapter 28

MATCHING

Match the description to the name of the physicists.

a) proposed the plum-pudding model
b) discovered radium and polonium
c) invented particle detector tube
d) observed back-scattering from nuclei
e) invented the bubble chamber
f) proposed the tiny central nucleus idea
g) discovered radioactivity

1. J. J. Thomson

   Answer: a) proposed the plum-pudding model   Difficulty: 2

2. Marie & Pierre Curie

   Answer: b) discovered radium and polonium   Difficulty: 2

3. Hans Geiger

   Answer: c) invented particle detector tube   Difficulty: 2

4. Ernest Marsden

   Answer: d) observed back-scattering from nuclei   Difficulty: 2

5. Donald Glazer

   Answer: e) invented the bubble chamber   Difficulty: 2

6. Ernest Rutherford

   Answer: f) proposed the tiny central nucleus idea   Difficulty: 2

7. Henri Becquerel

   Answer: g) discovered radioactivity   Difficulty: 2

Match the description to the unit.

a) SI unit of radioactivity (= 1 decay/s)
b) 10^{-2} J/kg
c) dosage of 2.58×10^{-4} C/kg
d) decay constant
e) 3.70×10^{10} decays/s in radium
f) SI unit of absorbed dose = 1 J/kg = 100 rad
g) effective dose (= (rads)(RBE))
h) effective dose of 1 Gy
i) produces same tissue damage as 1 rad

8. becquerel (Bq)

   Answer: a) SI unit of radioactivity (= 1 decay/s) Difficulty: 2

9. rad

   Answer: b) 10^{-2} J/kg Difficulty: 2

10. roentgen (R)

    Answer: c) dosage of 2.58×10^{-4} C/kg Difficulty: 2

11. s^{-1}

    Answer: d) decay constant Difficulty: 2

12. curie (Ci)

    Answer: e) 3.70×10^{10} decays/s in radium Difficulty: 2

13. gray (Gy)

    Answer: f) SI unit of absorbed dose = 1 J/kg = 100 rad Difficulty: 2

14. rem

    Answer: g) effective dose (= (rads)(RBE)) Difficulty: 2

15. sievert (Sv)

    Answer: h) effective dose of 1 Gy Difficulty: 2

16. RBE

    Answer: i) produces same tissue damage as 1 rad Difficulty: 2

Chapter 28

ESSAY

17. Write the nuclear notation for the following nuclei: Hydrogen-2, Sulfur-33, and Lead-207.

    Answer:
    $^{2}_{1}H$, $^{33}_{16}S$, $^{207}_{82}Pb$

    Difficulty: 2

MULTIPLE CHOICE

18. Which of the following statements concerning the nuclear force is false?
    a) The nuclear force is very short-ranged.
    b) The nuclear force is very weak and much smaller in relative magnitude than the electrostatic and gravitational forces.
    c) The nuclear force is attractive and not repulsive.
    d) The nuclear force acts on both protons and neutrons.

    Answer: b Difficulty: 2

19. Compared to the electrostatic force, the nuclear force between adjacent protons in a nucleus is
    a) much weaker.
    b) about the same size.
    c) only slightly larger.
    d) none of the above.

    Answer: d Difficulty: 2

20. If an element of atomic number 15 has an isotope of mass number 32,
    a) the number of neutrons in the nucleus is 15.
    b) the number of neutrons in the nucleus is 17.
    c) the number of protons in the nucleus is 17.
    d) the number of nucleons in the nucleus is 15.

    Answer: b Difficulty: 2

21. Isotopes of an element have nuclei with
    a) the same number of protons, but different numbers of neutrons.
    b) the same number of protons, and the same number of neutrons.
    c) a different number of protons, and a different number of neutrons.
    d) a different number of protons, and the same numbers of neutrons.

    Answer: a Difficulty: 2

TRUE/FALSE

22. When a nucleus undergoes beta decay, it changes into another isotope of itself.

    Answer: F Difficulty: 2

23. All elements have the same number of isotopes.

    Answer: F Difficulty: 2

24. Not all isotopes are radioactive.

    Answer: T Difficulty: 2

MULTIPLE CHOICE

25. The number of protons in an atom is
    a) zero.
    b) called the mass number.
    c) equal to the number of neutrons.
    d) equal to the number of electrons.
    e) the same for all elements.

    Answer: d Difficulty: 2

26. The mass of an atom is
    a) approximately equally divided between neutrons, protons, and electrons.
    b) evenly divided between the nucleus and the surrounding electron cloud.
    c) concentrated in the cloud of electrons surrounding the nucleus.
    d) concentrated in the nucleus.
    e) about 10^{-6} g.

    Answer: d Difficulty: 2

27. How many protons are there in the carbon-14 nucleus?
    a) None
    b) 1
    c) 6
    d) 8
    e) 14

    Answer: c Difficulty: 2

Chapter 28

28. Heavy-water molecules have a mass number of
    a) 3
    b) 10
    c) 12
    d) 18
    e) 20

    Answer: e Difficulty: 2

ESSAY

29. What is the approximate nuclear radius of an isotope of sodium with 11 protons and 12 neutrons?

    Answer: 3.4×10^{-15} m Difficulty: 2

MULTIPLE CHOICE

30. From a knowledge of the size and mass of a nucleus, one can estimate the density of nuclear material. Such a calculation shows that the density of nuclear matter is on the order of magnitude of
    a) 10^{17} kg/m^3
    b) 10^{11} kg/m^3
    c) 10^8 kg/m^3
    d) 10^4 kg/m^3
    e) 10^2 kg/m^3

    Answer: a Difficulty: 2

31. Atoms with the same atomic number but with different numbers of neutrons are referred to as
    a) nucleons.
    b) nuclides.
    c) isotopes.
    d) none of the above.

    Answer: a Difficulty: 2

32. An atom's mass number is determined by the number of
    a) neutrons in its nucleus.
    b) nucleons in its nucleus.
    c) protons in its nucleus.
    d) alpha particles in its nucleus.

    Answer: b Difficulty: 2

33. If an atom's atomic number is given by Z, its atomic mass by A, and its neutron number by N, which of the following is correct?
    a) N = A + Z
    b) N = Z - A
    c) N = A - Z
    d) None of the above is correct.

    Answer: c Difficulty: 2

34. There is a limit to the size of a stable nucleus because of
    a) the limit range of the strong nuclear force.
    b) the weakness of the electrostatic force.
    c) the weakness of the gravitational force.
    d) none of the above.

    Answer: a Difficulty: 2

35. When a gamma ray is emitted from an unstable nucleus,
    a) the number of neutrons and the number of protons drop by two.
    b) the number of neutrons drops by one and the number of protons increases by one.
    c) there is no change in either the number of neutrons or the number of protons.
    d) none of the above is correct.

    Answer: c Difficulty: 2

36. When an alpha particle is emitted from an unstable nucleus, the atomic number of the nucleus
    a) increases by 2.
    b) decreases by 2.
    c) increases by 4.
    d) decreases by 4.
    e) none of the above.

    Answer: b Difficulty: 2

37. When an alpha particle is emitted from an unstable nucleus, the atomic mass number of the nucleus
    a) increases by 2.
    b) decreases by 2.
    c) increases by 4.
    d) decreases by 4.
    e) none of the above.

    Answer: d Difficulty: 2

Chapter 28

38. When a β- particle is emitted from an unstable nucleus, the atomic number of the nucleus
   a) increases by 1.
   b) decreases by 1.
   c) does not change.
   d) none of the above.

   Answer: a   Difficulty: 2

39. When a β+ particle is emitted from an unstable nucleus, the atomic number of the nucleus
   a) increases by 1.
   b) decreases by 1.
   c) does not change.
   d) none of the above.

   Answer: b   Difficulty: 2

40. When a neutron is emitted from an unstable nucleus, the atomic mass number of the nucleus
   a) increases by 1.
   b) decreases by 1.
   c) does not change.
   d) none of the above.

   Answer: b   Difficulty: 2

41. During β- decay
   a) a neutron is transformed to a proton.
   b) a proton is transformed to a neutron.
   c) a neutron is ejected from the nucleus.
   d) a proton is ejected from the nucleus.

   Answer: a   Difficulty: 2

42. During β+ decay
   a) a neutron is transformed to a proton.
   b) a proton is transformed to a neutron.
   c) a neutron is ejected from the nucleus.
   d) a proton is ejected from the nucleus.

   Answer: b   Difficulty: 2

43. During K-capture
   a) a proton is absorbed by the nucleus.
   b) a neutron is absorbed by the nucleus.
   c) an electron is absorbed by the nucleus.
   d) an alpha particle is absorbed by the nucleus.

   Answer: c   Difficulty: 2

TRUE/FALSE

44. No charged particle is emitted as a result of K-capture.

    Answer: T Difficulty: 2

ESSAY

45. Complete the following nuclear decay equation: $^{210}_{84}Po^* \rightarrow {}^{210}_{84}Po$ + _____

    Answer: γ radiation Difficulty: 2

46. Complete the following nuclear decay equation: $^{238}_{92}U \rightarrow {}^{234}_{90}Th$ + _____

    Answer:
    $^{4}_{2}He$ (α particle)

    Difficulty: 2

47. Complete the following nuclear decay equation: $^{22}_{11}Na$ + _____ → $^{22}_{10}Ne$

    Answer:
    $^{1}_{0}n$ (neutron)

    Difficulty: 2

MULTIPLE CHOICE

48. The radioactivity due to carbon-14 measured in a piece of a wooden casket from an ancient burial site was found to produce 20 counts per minute from a given sample, whereas the same amount of carbon from a piece of living wood produced 160 counts per minute. The half-life of carbon-14, a beta emitter, is 5730 years. Thus we would estimate the age of the artifact to be about
    a) 5,700 years
    b) 11,500 years
    c) 14,800 years
    d) 17,200 years
    e) 23,000 years

    Answer: d Difficulty: 2

Chapter 28

ESSAY

49. A radioactive sample is determined to have 3.45×10^{19} nuclei present. Two days later (48 hours) it is again tested, but now found to have 2.00×10^{18} nuclei present. How many half-lives have elapsed during these 48 hours?

    Answer: 4.11   Difficulty: 2

50. Suppose that in a certain collection of nuclei there were initially 1024 nuclei and 20 minutes later there was only one nuclei left, the others having decayed. On the basis of this information, how many nuclei would you estimate decayed in the first 6 minutes?

    Answer: 896   Difficulty: 2

MULTIPLE CHOICE

51. A radioactive substance with a half-life of 3 days has an initial activity of 0.24 Ci. What is its activity after 6 days?
    a) 0.12 Ci
    b) 0.48 Ci
    c) 0.06 Ci
    d) none of the above

    Answer: c   Difficulty: 2

ESSAY

52. A radioactive substance containing 4.00×10^{16} unstable atoms has a half-life of 2 days. What is its activity (in curies) after 1 day?

    Answer: 3.07 Ci   Difficulty: 2

MULTIPLE CHOICE

53. In radioactive dating, carbon-14 is often used. This nucleus emits a single beta particle when it decays. When this happens, the resulting nucleus is
    a) still carbon-14.
    b) boron-14.
    c) nitrogen-14.
    d) carbon-15.
    e) carbon-13.

    Answer: c   Difficulty: 2

54. What happens to the half-life of a radioactive substance as it decays?
   a) It remains constant.
   b) It increases.
   c) It decreases.
   d) It could do any of these.

   Answer: a   Difficulty: 2

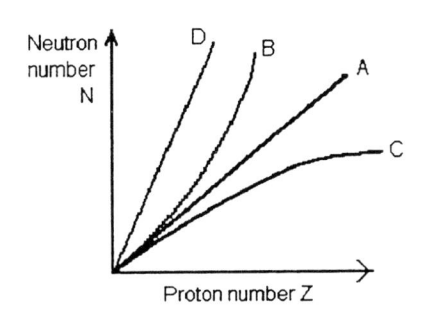

55. The neutron-proton stability curve of the isotopes of all elements will be plotted on a graph whose axes are shown. The plot will look like
   a) the straight line (N = Z), shown.
   b) a curve above the line for N > Z.
   c) a curve below the line for N < Z.
   d) a straight line where N > Z.

   Answer: b   Difficulty: 2

TRUE/FALSE

56. Stable nuclei with mass numbers greater than 40 have more neutrons than protons.

   Answer: T   Difficulty: 2

57. carbon-12 is unstable.

   Answer: F   Difficulty: 2

58. The half-life of carbon-14 is greater than the half-life of cobalt-60.

   Answer: T   Difficulty: 2

ESSAY

59. Calculate the binding energy of ^{9}Be.

   Answer: 58.2 MeV   Difficulty: 2

60. How much energy is required to remove one proton from $_9$Be?

   Answer: 16.9 MeV   Difficulty: 2

Chapter 28

MULTIPLE CHOICE

61. Which of the following combinations of nucleons would be the most likely to result in a stable nucleus? ("Odd" and "even" refer to the numbers of protons and neutrons in the nucleus.)
    a) Odd protons, even neutrons
    b) Even protons, odd neutrons
    c) Odd protons, odd neutrons
    d) Even protons, even neutrons

    Answer: d Difficulty: 2

62. What is the binding energy of the ^{4}He nucleus?
    a) 7.80 MeV
    b) 14.15 MeV
    c) 20.36 MeV
    d) 28.3 MeV

    Answer: d Difficulty: 2

63. The binding energy per nucleon
    a) increases steadily as we go to heavier elements.
    b) decreases steadily as we go to heavier elements.
    c) is approximately constant throughout the periodic table, except for very light nuclei.
    d) has a minimum near iron in the periodic table.

    Answer: d Difficulty: 2

64. The binding energy per nucleon is
    a) directly proportional to atomic number.
    b) inversely proportional to atomic number.
    c) the same for all atoms.
    d) none of the above.

    Answer: d Difficulty: 2

65. The type of detector that requires a magnetic field to "view" charged particles is a
    a) Geiger tube.
    b) scintillation counter.
    c) cloud chamber.
    d) bubble chamber.
    e) spark chamber.

    Answer: d Difficulty: 2

66. The type of detector that uses liquid hydrogen is a
   a) Geiger tube.
   b) scintillation counter.
   c) cloud chamber.
   d) bubble chamber.
   e) spark chamber.

   Answer: d Difficulty: 2

67. Cloud chambers have been replaced by bubble chambers because
   a) the radioactive clouds were too dangerous to work with.
   b) the density of fluids is greater than the density of vapors.
   c) bubble chambers tend to be larger and more expensive.
   d) gamma rays are visible in bubble chambers, but not in cloud chambers.

   Answer: b Difficulty: 2

TRUE/FALSE

68. The dead time of a Geiger counter is less than the dead time of a scintillation counter.

   Answer: F Difficulty: 2

MULTIPLE CHOICE

69. The chief hazard of radiation is
   a) damage to living cells due to ionization.
   b) damage to cells due to heating.
   c) damage to living cells due to the creation of chemical impurities.
   d) the creation of new isotopes within the body.

   Answer: a Difficulty: 2

70. A unit that measures the effective dose of radiation in a human is the
   a) curie.
   b) RBE.
   c) rad.
   d) rem.

   Answer: d Difficulty: 2

71. All of the following are units used to describe radiation dosage in humans except
   a) curie.
   b) rad.
   c) rem.
   d) RBE.
   e) sievert.

   Answer: a Difficulty: 2

Chapter 28

72. In which of the following applications of technology in the home is radioactivity utilized?
   a) A door-bell
   b) A humidifier
   c) A smoke detector
   d) A food sterilizer

   Answer: c Difficulty: 2

73. The presence and magnitude of corrosion in steel pipes is commonly detected by using
   a) QF measurements.
   b) nuclear magnetic resonance.
   c) computer-assisted tomography.
   d) neutron activation analysis.

   Answer: b Difficulty: 2

74. Assuming the Earth's orbital angular momentum is quantized, estimate its angular momentum quantum number (l).
   a) 3×10^{24}
   b) 3×10^{44}
   c) 3×10^{74}
   d) 3×10^{-34}

   Answer: c Difficulty: 2

FILL-IN-THE-BLANK

75. Assuming you are particle in a box called the universe. Estimate your minimum kinetic energy (Assume the "edge" of the box is at a distance of 20.Giga Light Years).

   Answer: 6×10^{-123} Joules = 3×10^{-104} eV from $\lambda/2 = 2R$ (R=20GLY) and $K = p^2/2m = h^2/(2$

   Difficulty: 2

TRUE/FALSE

76. The 4d subshell can contain 14 electrons.

   Answer: T Difficulty: 1

77. The 5d subshell can contain more electrons than the 3d subshell.

   Answer: F Difficulty: 1

MULTIPLE CHOICE

78. In a multielectron atom, which of the following levels is just above the 6s level?
    a) 7s
    b) 6p
    c) 5p
    d) 4f
    e) 4d

    Answer: d Difficulty: 2

FILL-IN-THE-BLANK

79. The electron spin quantum number can take on what values?

    Answer: +1/2 and -1/2 Difficulty: 2

80. For an orbital quantum number of 2, the magnetic quantum number can take on what allowed values?

    Answer: 2, 1, 0, -1, -2 Difficulty: 2

MULTIPLE CHOICE

81. Who in 1926 developed the general wave equation for the de Broglie matter waves?
    a) Heisenberg
    b) Pauli
    c) Einstein
    d) Schrodinger
    e) Bohr

    Answer: d Difficulty: 1

ESSAY

82. Explain the allowed values of the four quantum numbers for the hydrogen atom.

    Answer:
    Principle quantum number: $n = 1,2,3,4,.....$
    Orbital number: $l = 0,1,2,....,(n-1)$
    Orbital magnetic number: $|m| = 0,1,2,...,l$
    Spin quantum number: $|m_s| = 1/2$

    Difficulty: 2

Chapter 28

FILL-IN-THE-BLANK

83. In the periodic table, the horizontal rows are called ____ and the vertical columns are called ____.

Answer: periods, groups   Difficulty: 2

MULTIPLE CHOICE

84. Who suggested that material particles (protons, etc) have wave properties too?
    a) Einstein
    b) Bohr
    c) Planck
    d) Schrodinger
    e) de Broglie

Answer: e Difficulty: 2

85. Natural line broadening can be understood in terms of the:
    a) de Broglie wave length
    b) uncertainty principle
    c) Schrodinger wave equation
    d) Pauli exclusion principle

Answer: b Difficulty: 2

ESSAY

86. Compare the prediction of the Bohr theory with the Schrodinger theory of the hydrogen atom concerning the angular momentum of the ground state (similarity or difference).

Answer:
Both theories assumed quantized angular momentum, however the Bohr theory was based on $L=h/(2\pi)$ for the ground state, whereas Schrodinger's wave equation predicts ZERO angular momentum in this lowest state.
Difficulty: 2

MULTIPLE CHOICE

87. In the periodic table of elements, members of each ___ have similar chemical properties.
    a) period
    b) group
    c) transition
    d) row

Answer: b Difficulty: 2

88. Heisenberg is especially noted for his:
    a) photoelectric theory
    b) improved periodic table
    c) dirigible
    d) wave equation
    e) uncertainty principle

    Answer: e Difficulty: 2

89. The third SHELL has how many sub-shells?
    a) 1
    b) 2
    c) 3
    d) 4
    e) 5
    f) 6

    Answer: e Difficulty: 2

ESSAY

90. If someone were to suggest that possibly half of the galaxies in the universe are composed of antiparticles ("anti-galaxies"), what argument(s) might you present to refute this hypothesis? Remember, atoms made of antiparticles have the same chemical and spectroscopic properties that our "ordinary" matter has.

    Answer:
    Astronomers see many galaxies colliding and if a pair were composed of a galaxy and an antigalaxy, then ENORMOUS high energy radiation would escape (be seen) due to the particle-antiparticle annihilations.
    Difficulty: 2

FILL-IN-THE-BLANK

91. A 20.gram marble is trapped in a 10.cm box (1-dim) with an energy of 0.10 Joule.
    a) estimate the quantum number for this level.
    b) what is the energy of the ground state?

    Answer: a) $n = 1.9 \times 10^{31}$ b) 2.7×10^{-64} J = 1.7×10^{-45} eV

    Difficulty: 2

Chapter 28

ESSAY

92. Discuss what makes the Schrodinger wave function so non-classical.

   Answer:
   Although the wave function fully describes what can be known about a particle, it only makes probability statements. Its square determines the probability density of location................
   Difficulty: 2

MULTIPLE CHOICE

93. What is the momentum of a 600.nm photon?
   a) 2.07 N-s
   b) 2.07 eV/c
   c) 2.07 KeV/c
   d) 2.07 MeV/c

   Answer: b Difficulty: 2

Chapter 29

ESSAY

1. Complete the following nuclear reaction:

   $^{16}_{8}O + ^{1}_{0}n \rightarrow \underline{} + ^{4}_{2}He$

   Answer:
   $^{13}_{6}C*$

   Difficulty: 2

2. Complete the following nuclear reaction:

   $^{27}_{13}Al\ (\alpha,n)\ \underline{}$

   Answer:
   $^{31}_{15}P*$

   Difficulty: 2

TRUE/FALSE

3. A negative Q value for a reaction indicates the reaction is exoergic.

   Answer: F Difficulty: 2

ESSAY

4. Find the Q value of the following reaction:

   $\quad ^{14}_{7}N \quad + \quad ^{4}_{2}He \quad \rightarrow \quad ^{17}_{8}O \quad + \quad ^{1}_{1}H$
   (14.003074 u) (4.002603 u) (16.999131 u) (1.007825 u)

   Answer: -0.001279 u Difficulty: 2

5. A carbon-14 nucleus decays to a nitrogen-14 nucleus by beta decay. How much energy (in MeV) is released if carbon-14 has a mass of 14.003074 u and nitrogen-14 has a mass of 14.003242 u?

   Answer: 0.157 MeV Difficulty: 2

Chapter 29

MULTIPLE CHOICE

6. The mass of a proton is 1.6726×10^{-27} kg and the mass of a neutron is 1.6606×10^{-27} kg. A proton captures a neutron forming a deuterium nucleus. One would expect the mass of this nucleus to be
   a) equal to $(1.6726 + 1.6606) \times 10^{-27}$ kg.
   b) less than $(1.6726 + 1.6606) \times 10^{-27}$ kg.
   c) greater than $(1.6726 + 1.6606) \times 10^{-27}$ kg.
   d) any of these, it depends on the energy released during the capture.

   Answer: b Difficulty: 2

7. The expression $(M_X - M_Y - M_a) \times 931.5$ represents
   a) the binding energy of the nucleus X.
   b) the binding energy of the nucleus Y.
   c) the energy released when nucleus X undergoes alpha decay.
   d) the energy released when nucleus Y undergoes alpha decay.
   e) the energy released in a typical fission reaction of uranium.

   Answer: c Difficulty: 2

ESSAY

8. Find the energy in MeV released in the reaction

   $^{238}_{92}U \rightarrow \, ^{234}_{90}Th + \, ^{4}_{2}He$

   Mass of U-238 = 238.050786 u
   Mass of Th-234 = 234.043583 u
   Mass of He-4 = 4.002603 u

   Answer: 4.28 MeV Difficulty: 2

MULTIPLE CHOICE

9. The Q value for a particular reaction is -2.4 MeV, and the reaction's threshold energy is 9.60 MeV. What is the ratio of the mass of the incident particle to the mass of the stationary target nucleus?
   a) -0.75
   b) 0.25
   c) 3
   d) 4
   e) 5

   Answer: c Difficulty: 2

10. When a target nucleus is bombarded by an appropriate beam of particles, it is possible to produce
    a) a less massive nucleus, but not a more massive one.
    b) a more massive nucleus, but not a less massive one.
    c) a nucleus with smaller atomic number, but not one with a greater atomic number.
    d) a nucleus with greater atomic number, but not one with a smaller atomic number.
    e) a nucleus with either greater or smaller atomic number.

    Answer: e Difficulty: 2

11. What is the energy of reaction of the process $^9_4Be(\alpha,n)^{12}_6C$?

    (This reaction, first observed by Chadwick in 1930, led to his discovery of the neutron.)
    a) 3.66 Mev
    b) 5.60 MeV
    c) 5.70 MeV
    d) 6.11 MeV
    e) 6.34 MeV

    Answer: c Difficulty: 2

12. Which of the following is most nearly the same as a gamma ray?
    a) An alpha particle
    b) A beta ray
    c) Visible light
    d) A proton
    e) A neutron

    Answer: c Difficulty: 2

13. What is the source of the energy the sun radiates to us?
    a) Chemical reactions
    b) Nuclear fission reactions
    c) Nuclear fusion reactions
    d) Magnetic explosions
    e) Cosmic rays

    Answer: c Difficulty: 2

14. A proton strikes an oxygen-18 nucleus producing fluorine-18 and another particle. What other particle is produced by this nuclear reaction?
    a) A neutron
    b) An alpha particle
    c) A β- particle
    d) A β+ particle

    Answer: a Difficulty: 2

Chapter 29

15. Which of the following best describes the process in which energy is released in a conventional nuclear reactor?
   a) The radiation given off by a naturally radioactive substance, uranium, is collected and used to make steam.
   b) Uranium is reacted with oxygen in a combustion process that releases large amounts of radioactivity and heat.
   c) Deuterium and tritium are joined together to form helium.
   d) Uranium, when bombarded by neutrons, splits into fragments and releases two or three neutrons, and these neutrons in turn strike more uranium nuclei that split, thereby setting off a chain reaction that releases energy.
   e) A uranium nucleus is energized to an excited state by neutron irradiation, and it then decays by emitting beta rays and gamma rays that heat water and create steam.

   Answer: d   Difficulty: 2

16. Which of the following (if any) statements is true concerning a nuclear reactor as compared to a coal-fired power plant for generating electricity?
   a) Higher voltages are generated when nuclear reactions are used.
   b) The electricity from a nuclear reactor is slightly radioactive, whereas that from a coal-fired plant is not.
   c) Nuclear energy can be converted directly into electricity, whereas a coal-fired plant must first generate steam, which in turn drives a turbine.
   d) The radioactivity in the smoke from a coal-fired plant is comparable to that from a nuclear plant, but since it is so widely dispersed it does not present a major environmental hazard.
   e) None of these statements is true.

   Answer: e   Difficulty: 2

17. When light elements such as hydrogen undergo fusion,
   a) there is a loss of mass.
   b) there is an increase in mass.
   c) there is no change in mass.
   d) electric charge can be annihilated.
   e) more than one of these statements is true.

   Answer: a   Difficulty: 2

18. In the fission reaction $^{235}U + {}^1n \rightarrow {}^{141}Ba + {}^{92}Kr$ + neutrons, the number of neutrons produced is
   a) zero.
   b) 1.
   c) 2.
   d) 3.

   Answer: d   Difficulty: 2

19. In a nuclear power reactor of the type used to generate electricity, a neutron bombards a uranium nucleus, causing it to split into two large pieces (fission fragments) plus two or three neutrons. Energy is released in this process in the form of electromagnetic radiation and kinetic energy. Which of the following is an accurate statement concerning what happens in such a fission process?
   a) The electrical energy generated comes from the kinetic energy of the incident neutrons.
   b) The electrical energy generated comes from chemical energy stored in the electron bonds of the uranium atom.
   c) Electrical energy is generated because the fission fragments are electrically charged, whereas the uranium was electrically neutral.
   d) An intermediate step involves the fusion of protons to form helium nuclei, and energy is released in this process.
   e) The total mass of the particles after fission is less than the total mass of the particles (uranium nucleus plus one neutron) before fission, and this decrease in mass $_0 m$ is converted into energy E, where $E = \Delta mc^2$.

   Answer: e Difficulty: 2

20. What is the principle difference between a hydrogen bomb and a uranium bomb?
   a) A uranium bomb is an atomic bomb, and a hydrogen bomb is a nuclear bomb.
   b) A uranium bomb utilizes a fission reaction whereas a hydrogen bomb utilizes a fusion reaction.
   c) Both work on the same principle, but the hydrogen bomb has a higher yield.
   d) A hydrogen bomb converts mass into energy, whereas a uranium bomb does not.
   e) One results in radioactive fallout, and the other does not.

   Answer: b Difficulty: 2

21. One sometimes hears reference to a "20-kiloton" bomb. What does this mean?
   a) It means 20,000 tons of nuclear explosive is used in the bomb.
   b) It means that the number of "fissile" nuclei is equal to the number of trinitrotoluene molecules in 20,000 tons of TNT explosive.
   c) It means that the total weight of the bomb (not just the uranium) is 20,000 tons.
   d) It means that the energy released by the bomb is equal to the energy releasedwhen 20,000 tons of TNT is exploded.
   e) This refers to the maximum pressure generated by the bomb when it explodes.

   Answer: d Difficulty: 2

Chapter 29

22. What is the meaning of the term "critical mass"?
    a) This refers to the mass of the "critical" elements in a reactor, i.e., the uranium or plutonium.
    b) This refers to the minimum amount of fissionable material required to sustain a chain reaction.
    c) This is the amount of mass needed to make a power reactor economically feasible.
    d) This is the material which is just on the verge of becoming radioactive.

    Answer: b Difficulty: 2

23. The energy radiated by a star, such as the sun, results from
    a) beta decay.
    b) alpha decay.
    c) fission reactions.
    d) fusion reactions.

    Answer: d Difficulty: 2

24. The neutrino has
    a) enormous rest mass, positive charge, and spin quantum number 1/2.
    b) negligible rest mass, negative charge, and spin quantum number 1.
    c) negligible rest mass, no charge, and spin quantum number 1/2.
    d) none of the above.

    Answer: c Difficulty: 2

25. The existence of the neutrino was postulated in order to explain
    a) alpha decay.
    b) gamma emission.
    c) beta decay.
    d) fission.

    Answer: c Difficulty: 2

26. Scientists were led to postulate the existence of the neutrino in order to
    a) maintain the conservation of energy and of momentum in beta decay.
    b) explain the intense radiation emitted by quasars.
    c) account for alpha decay.
    d) provide a mechanism for quark production.

    Answer: a Difficulty: 2

27. In beta decay
    a) a proton is emitted.
    b) a neutron is emitted.
    c) an electron is emitted.
    d) an electron decays into another particle.

    Answer: c Difficulty: 2

28. Which of the following is not considered to be one of the four fundamental forces?
a) The gravity force
b) The meson force
c) The weak nuclear force
d) The strong nuclear force
e) The electromagnetic force

Answer: b Difficulty: 2

29. What effect does an increase in the mass of the virtual exchange particle have on the range of the force it mediates?
a) Decreases it.
b) Increases it.
c) Has no appreciable effect.
d) Decreases charged particle interactions and increases neutral particle interactions.

Answer: a Difficulty: 2

30. A particle that travels at the speed of light
a) has never been observed.
b) must have zero rest mass.
c) must have a very large rest mass.
d) has infinite energy.

Answer: b Difficulty: 2

31. The exchange particles for the weak force are very massive (about 100 times as massive as a proton). This would lead one to expect that the weak force would
a) act over a very long range.
b) act over a very short range.
c) only act on very massive particles.
d) be transmitted at the speed of light.

Answer: b Difficulty: 2

32. An electron is an example of
a) a hadron.
b) a meson.
c) a lepton.
d) a baryon.

Answer: c Difficulty: 2

Chapter 29

33. The exchange particle for quarks is called
    a) the stickon.
    b) the gluon.
    c) the epoxyon.
    d) the epsilon.

    Answer: b Difficulty: 2

34. The exchange particle for the weak force is the
    a) photon.
    b) meson.
    c) W.
    d) graviton.

    Answer: c Difficulty: 2

35. Which of the following is true?
    a) All hadrons are baryons or leptons.
    b) All hadrons are leptons or mesons.
    c) All hadrons are mesons or baryons.
    d) All hadrons are nucleons.

    Answer: c Difficulty: 2

TRUE/FALSE

36. Nucleons are baryons, not hadrons.

    Answer: T Difficulty: 2

MULTIPLE CHOICE

37. A distinctive feature of quarks is that they
    a) have zero rest mass.
    b) have zero charge.
    c) have fractional electric charge.
    d) are always observed singly, since they do not readily interact with other particles.

    Answer: c Difficulty: 2

38. A significant step in "unifying" the forces of nature was the discovery that two of the so-called fundamental forces were two parts of a single force. The two forces that were so unified were
    a) the electromagnetic force and the weak force.
    b) the electromagnetic force and the strong nuclear force.
    c) the electromagnetic force and the gravity force.
    d) the weak force and the strong nuclear force.

    Answer: a Difficulty: 2

39. The grand unified theory (GUT) is a sought-after model that would unify three of the four forces of nature. Which force is the one it would not include?
   a) Gravitational
   b) Strong
   c) Weak
   d) Electromagnetic

   Answer: a Difficulty: 2

40. A quark is
   a) a constituent of a nucleon.
   b) a constituent of a hadron.
   c) an elementary particle.
   d) all of the above.

   Answer: d Difficulty: 2

TRUE/FALSE

41. Mesons consist of quark-antiquark pairs.

   Answer: T Difficulty: 2

MULTIPLE CHOICE

42. The number of quarks in a deuteron (^2_1H) is

   a) 1
   b) 2
   c) 3
   d) 4
   e) 6

   Answer: e Difficulty: 2

Chapter 29

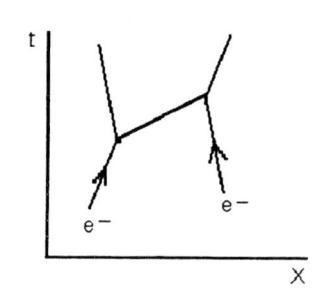

43. The Feynman diagram shows two electrons approaching each other, interacting, then leaving each other. What particle is being exchanged during the interaction?
a) Pion
b) Virtual photon
c) Neutrino
d) W particle

Answer: b Difficulty: 2

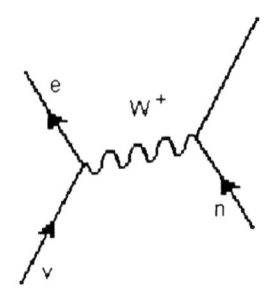

44. The Feynman diagram shows the weak interaction of a neutrino (v) and a neutron (n), mediated by a W^+ particle. The interaction produces an electron (e^-) and another particle. What is the other particle?
a) A positron (e^+).
b) A pion (π^+).
c) A proton (p).
d) A quark.
e) An anti-neutron.

Answer: c Difficulty: 2

45. The number of types of quarks (including the "top" quark, which has not yet been observed) is
a) 2.
b) 3.
c) 6.
d) 8.

Answer: c Difficulty: 2

46. When a quark emits or absorbs a gluon, the quark changes
    a) its charge.
    b) its mass.
    c) its color.
    d) into an antiquark.

    Answer: c   Difficulty: 2

TRUE/FALSE

47. Carbon-14 dating can reliably date ancient archeological samples up to approximately 1000. million years.

    Answer: F   Difficulty: 1

48. Unstable nuclei that undergo positron decay have more neutrons than protons.

    Answer: F   Difficulty: 1

MULTIPLE CHOICE

49. Which of the following have the greatest RBE (relative biological effectiveness)?
    a) slow neutrons
    b) beta particles
    c) fast neutrons
    d) alpha particles
    e) gamma rays

    Answer: d   Difficulty: 2

FILL-IN-THE-BLANK

50. The normal U.S. average annual radiation dose per person is about how many REM?
    _____

    Answer: 0.20   Difficulty: 2

MULTIPLE CHOICE

51. A nucleus which undergoes alpha decay:
    a) increases its proton number
    b) increases its neutron number
    c) increases its electron number
    d) decreases its proton number

    Answer: d   Difficulty: 1

Chapter 29

52. A nucleus which undergoes beta minus decay undergoes an:
    a) increase of neutron number
    b) increase of proton number
    c) decrease of proton number
    d) none of these

    Answer: b Difficulty: 1

53. All isotopes are unstable which have proton number greater than:
    a) 73
    b) 78
    c) 83
    d) 4
    e) 12
    f) 26

    Answer: c Difficulty: 2

54. Plutonium has a half life of 2.4×10^4 years. How long does it take for 99% of the Plutonium to decay?
    a) 1.6×10^6 years
    b) 1.6×10^5 y
    c) 1.6×10^4 y
    d) 1.6×10^3 y

    Answer: b Difficulty: 2

ESSAY

55. Why do the high-proton-number elements tend to get more unstable?

    Answer:
    The repulsive elctrostatic force between protons increases progressively as atomic number increases but the nuclear attractive force is short range and the average distance between particles increases as the nucleus size increases.
    Difficulty: 2

FILL-IN-THE-BLANK

56. Polonium-215 alpha decays with a half life of 1.8 ms.
    a) How much is left after 1.second?
    b) What is the daughter nucleus?

    Answer: a) fraction remaining: $N/N_o = 4.2 \times 10^{-170}$ b) Lead-213

    Difficulty: 2

MULTIPLE CHOICE

57. Which isotope is used to define the ATOMIC MASS UNIT ?
    a) ^1H
    b) ^2H
    c) ^4He
    d) ^{12}C
    e) ^{16}O

    Answer: b Difficulty: 1

58. The binding energy of a ^{170}Yb would be expected to be close to
    a) 8. MeV
    b) 170. MeV
    c) 1.4 GeV
    d) 170. GeV

    Answer: c Difficulty: 2

FILL-IN-THE-BLANK

59. Name the first three isotopes of hydrogen, their nuclide symbols, and indicate which is most common and which is most unstable.

    Answer:
    ordinary hydrogen ^1H, deuterium ^2H, and tritium ^3H ; ^1H being most common and ^3H decaying with a 12.year half life.

    Difficulty: 2

ESSAY

60. Discuss what observations convinced Rutherford that the nucleus was very small and dense.

    Answer:
    Alpha scattering from heavy nuclei produced much more backscattering than would be expected from a diffuse nuclear cloud. Only a dense compact nucleus of positive charge could account for the scattering at large angles.

    Difficulty: 2

Chapter 29

MULTIPLE CHOICE

61. Cosmic rays react with what atmospheric atoms to produce Carbon-14?
    a) Oxygen
    b) Hydrogen
    c) Nitrogen
    d) Carbon-12
    e) Helium

    Answer: c Difficulty: 2

62. The reciprocal of the decay constant is always _____ the half life.
    a) greater than
    b) equal to
    c) less than
    d) none of the above

    Answer: a Difficulty: 1

FILL-IN-THE-BLANK

63. Name five different types of radiation detectors.

    Answer:
    Geiger counter, scintillation counter, solid state detector, bubble chamber, spark chamber.
    Difficulty: 2

MULTIPLE CHOICE

64. Which of the following is NOT a unit of radioactive ACTIVITY:
    a) Becquerel
    b) rad
    c) Curie
    d) decays/s

    Answer: b Difficulty: 2

65. What is the threshhold level for radiation damage to humans?
    a) 0 rad
    b) 0.1 mrad
    c) 0.1 rad
    d) 5. rad
    e) 500. rad

    Answer: a Difficulty: 1

ESSAY

66. If the volume of the radius is approximately proportional to Mass Number, what does this say about nuclear densities?

    Answer:
    Nuclear mass (M) is closely proportional to mass number (A) so nuclear density = M/V proportional to A/A = constant. Hence nuclear densities are very similar (nucleus is incompressible as extra nucleons are added.)

    Difficulty: 2

Chapter 30

FILL-IN-THE-BLANK

1. Name the four known fundamental forces in nature:

   _____ _____ _____ _____

   Answer: Gravitation, Electromagnetic, Strong (Nuclear), Weak

2. What is the heaviest stable isotope?

   Answer: ^{209}Bi

MULTIPLE CHOICE

3. The proton is believed to consist of what quark combination? (up, down, strange, charm, top, bottom)
   a) uud
   b) ddu
   c) uds
   d) ttb
   e) bbt
   f) css

   Answer: a

4. Which one of the following particles can interact be all 4 of the known forces?
   a) muon
   b) anti-proton
   c) electron
   d) neutrino
   e) graviton

   Answer: b

FILL-IN-THE-BLANK

5. Concerning Uranium isotopes:
   a) Which is the most commonly occurring isotope?
   b) Which has a large cross section for slow neutrons?
   c) The majority of nuclear power plants depend upon which for fuel?

   Answer: a) ^{238}U   b) ^{235}U   c) ^{235}U

Testbank

MULTIPLE CHOICE

6. Which of the following is not considered a fundamental particle?
   a) alpha particle
   b) beta particle
   c) muon
   d) neutrino

   Answer: a

7. Which of the following is not believed to be composed of quarks?
   a) alpha particle
   b) beta particle
   c) proton
   d) neutron
   e) pion

   Answer: b

8. The mass of a neutron is _____ the combined mass of a proton and an electron.
   a) more than
   b) less than
   c) equal to

   Answer: a

FILL-IN-THE-BLANK

9. Complete the following reaction: $^1n + {}^{10}B \rightarrow {}^4He + ?$

   Answer: ^7Li

MULTIPLE CHOICE

10. The sun's energy comes from:
    a) nuclear fission
    b) nuclear fusion
    c) nuclear annihilation
    d) nuclear combustion
    e) alpha decay

    Answer: b

11. Which of the following is not massless?
    a) neutrino
    b) graviton
    c) photon
    d) muon
    e) all of the above

    Answer: d

Chapter 30

FILL-IN-THE-BLANK

12. In references to nuclear chain reactions, what is meant by the term CRITICAL MASS?

    Answer:
    Critical mass is the minimum mass of fissionable material necessary to produce a chain reaction.

ESSAY

13. Discuss why it is impossible for a particle to decay into a single particle of less mass.

    Answer:
    Observe the decay from the rest frame of the original particle, in which momentum is obviously zero. If a single particle were created that has less mass, then some of the original mass became kinetic energy and the final particle would be moving and have momentum. But momentum must be conserved and the final can not have momentum.

FILL-IN-THE-BLANK

14. Determine the missing reaction product
    $^1n + {}^{235}U = {}^{141}Ba + \_?\_ + 3\,({}^1n)$

    Answer: ^{92}Kr

15. Starting with ^{235}U, assume the following sequence of decays occur. Determine the correct isotopic product after each process has occurred:
    1) alpha decay
    2) beta minus decay
    3) alpha decay
    4) alpha decay
    5) beta minus decay
    6) alpha decay

    Answer:
    1) Thorium ^{231}Th 2) Protactinium ^{231}Pa 3) Actinium ^{227}Ac
    4) Fracium ^{223}Fr 5) Radium ^{223}Ra 6) Radon ^{219}Rn

MULTIPLE CHOICE

16. A nuclear reaction starts with 1.71 MeV of kinetic energy and a rest energy of 28.2 GeV. If the reaction releases 1.10 MeV, what was the Q of the reaction?
    a) 0.60 MeV
    b) 1.10 MeV
    c) 28.8 GeV
    d) none of these

    Answer: b

17. What kind of reactor produces more fissionable fuel than it consumes?
    a) LOCA
    b) moderated
    c) heavy water
    d) breeder
    e) none of the above

    Answer: d

FILL-IN-THE-BLANK

18. Identify with an H or a L to which family each of the following particles belong (Hadron or Lepton):
    neutron_____   electron_____   muon_____   proton_____
    neutrino_____ pion_____   tauon_____

    Answer:
    H L L H
    L H L

19. Name the exchange particles for each of the 4 "fundamental" forces.

    Answer:
    Strong: Pion; Electromagnetic: Photon;
    Weak: W and Z; Gravity: Graviton

ESSAY

20. Discuss the problems involved in achieving controlled fusion as a source of energy.

    Answer:
    The fusion of light elements requires a temperature of millions of degrees to overcome the Coulomb repulsion. Magnetic and inertial confinement are being developed to achieve the densities necessary for net energy release........